Education, Participatory Action Research, and Social Change

Education, Participatory Action Research, and Social Change

International Perspectives

Edited by
Dip Kapoor and
Steven Jordan

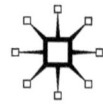

EDUCATION, PARTICIPATORY ACTION RESEARCH, AND SOCIAL CHANGE
Copyright © Dip Kapoor and Steven Jordan, 2009.
Softcover reprint of the hardcover 1st edition 2009 978-0-230-61513-7

All rights reserved.

First published in 2009 by
PALGRAVE MACMILLAN®
in the United States—a division of St. Martin's Press LLC,
175 Fifth Avenue, New York, NY 10010.

Where this book is distributed in the UK, Europe and the rest of the world, this is by Palgrave Macmillan, a division of Macmillan Publishers Limited, registered in England, company number 785998, of Houndmills, Basingstoke, Hampshire RG21 6XS.

Palgrave Macmillan is the global academic imprint of the above companies and has companies and representatives throughout the world.

Palgrave® and Macmillan® are registered trademarks in the United States, the United Kingdom, Europe and other countries.

ISBN 978-1-349-37883-8   ISBN 978-0-230-10064-0 (eBook)
DOI 10.1057/9780230100640

Library of Congress Cataloging-in-Publication Data
    International perspectives on education, PAR, and social change / edited by Dip Kapoor and Steven Jordan.
       p. cm.
    Includes bibliographical references and index.

     1. Action research—Case studies. 2. Social change—Case studies. I. Kapoor, Dip. II. Jordan, Steven.
H62.I6533 2009
303.4072—dc22   2009006635

A catalogue record of the book is available from the British Library.

Design by Newgen Imaging Systems (P) Ltd., Chennai, India.

First edition: October 2009

10 9 8 7 6 5 4 3 2 1

Transferred to Digital Printing in 2012

In fond memory of my father, Satya Kapoor—I dedicate this work to your unflagging sense of integrity and your constant compassion for those who are compelled to persevere.

—Dip

In memory of my brother, Roger David Jordan. A good friend whom I will always miss.

—Steve

CONTENTS

Acknowledgments ix

One   Introduction: International Perspectives on Education, PAR, and Social Change   1
Dip Kapoor & Steven Jordan

PART I   International Perspectives on Education and PAR

Two   From a Methodology of the Margins to Neoliberal Appropriation and Beyond: The Lineages of PAR   15
Steven Jordan

Three   Participatory academic research (*par*) and People's Participatory Action Research (PAR): Research, Politicization, and Subaltern Social Movements in India   29
Dip Kapoor

Four   When Research Becomes a Revolution: Participatory Action Research with Indigenous Peoples   45
Cora Weber-Pillwax

Five   *Ko tātou te rangahau, ko te rangahau, ko tātou*: A Māori Approach to Participatory Action Research   59
Lynne Harata Te Aika & Janinka Greenwood

Six   Translating "Participation" from North to South: A Case Against Intellectual Imperialism in Social Science Research   73
Cynthia M. Chambers & Helen Balanoff

Seven   Action Research for Curriculum Internationalization: Education versus Commercialization   89
Robin McTaggart & Gina Curró

Eight   Critical Complexity and Participatory Action Research: Decolonizing "Democratic" Knowledge Production   107
Joe L. Kincheloe

Nine	Reconceptualizing Participatory Action Research for Sustainability Education *Elizabeth A. Lange*	123

PART II   International Contexts: Case Studies of PAR, Education, and Social Change

Ten	*Chara chimwe hachitswanyi inda*: Indigenizing Science Education in Zimbabwe *Edward Shizha*	139
Eleven	Research and Agency: The Case of Rural Women and Land Tenure in Tanzania *Christine Hellen Mhina*	155
Twelve	NGO-Community Partnerships, PAR, and Learning in Mining Struggles in Ghana *Valerie Kwai Pun*	169
Thirteen	Ethnography-in-Motion: Neoliberalism and the Shack Dwellers Movement in South Africa *Shannon Walsh*	181
Fourteen	Kabyle Community Participatory Action Research (CPAR) in Algeria: Reflections on Research, *Amazigh* Identity, and Schooling *Taieb Belkacem*	195
Fifteen	Notes and Queries for an Activist Street Anthropology: Street Resistance, *Gringopolítica*, and the Quest for Subaltern Visions in Salvador da Bahia, Brazil *Samuel Veissière*	209
Sixteen	A Participatory Research Approach to Exploring Social Movement Learning in the Chilean Women's Movement *Donna M. Chovanec & Héctor M. González*	223
Seventeen	Participatory Research, NGOs, and Grassroots Development: Challenges in Rural Bangladesh *Bijoy P. Barua*	239
Eighteen	Making Space for Youth: iHuman Youth Society and Arts-Based Participatory Research with Street-Involved Youth in Canada *Diane Conrad & Wallis Kendal*	251

Contributors	265
Index	271

ACKNOWLEDGMENTS

An undertaking of this nature takes many minds and both of us would like to thank all the collaborators who have contributed toward developing the content of this collection. If we ever decide to put together another book, we would be more than happy to work with all of you again as you have been enthused about the idea from the first day and have replied to e-mails, editorial comments, and critique promptly and with a sense of dedication to your respective pieces and the project as a whole. Thank you all from the bottom of our minds.

Alison Crump has been painstakingly attentive in helping us edit the collection and to keep us thinking about questions of presentation and logic, if not the nitty-gritty details around referencing and formatting—Alison, your services were much appreciated and we look forward to working with you again. Special thanks to Kathy McElroy for her tireless, but always cheerful, help in the final stages of preparing the manuscript.

Immense gratitude and appreciation to the numerous communities that contributors to this collection have written for and with—without you, such a work would not have been conceivable—we hope this proves to be a small and useful contribution in your continuing journeys.

Thanks to Julia Cohen, our editor at Palgrave Macmillan who showed confidence in us and in the idea of this collection. We are also grateful to the anonymous external reviewer who provided us with a thought-provoking appraisal that has, hopefully, been responsible for making this a stronger and more worthwhile contribution.

Friends and family are mentioned last but are always first in our minds. Steven Jordan thanks Aziz Choudry for his many references and insights on the politics of PAR and his family, Elizabeth and Saoirse, whose support and patience were invaluable in getting the writing for this project completed. Dip Kapoor thanks his family, Paula, Niabi and Ahawi, for tolerating his absences, even when supposedly present at the dinner table and for their supportive good humor and constant love and care without which, life would be much harder to take.

Lastly, we would like to acknowledge the Social Sciences and Humanities Research Council of Canada (SSHRC) for their generous financial support in aid of this project.

CHAPTER ONE

Introduction: International Perspectives on Education, PAR, and Social Change

DIP KAPOOR & STEVEN JORDAN

As globalist euphoria, market triumphalism despite recent vulnerabilities and hypermodernizing political, cultural, educational, and linguistic projects define the dominant accents of this historical juncture, academics and knowledge/pedagogical workers continue to deliberate on our contributions to education, research, and to knowledge processes. More importantly, for those of us who view such engagements as being *fundamentally political*, we *critically assess* the import of synergies between research, knowledge, education, and social action in relation to the dominant epistemic conceptions and political-economic interests of our time. Recognizing that all research and socioeducational inquiry begins from a social location and that research being conducted by those who are located *with and for* power, will produce knowledge useful for the purposes of ruling relations and the reinforcement of existing hierarchies of culture and material existence (for instance, "western empiricism reifies the conventional values legitimating capitalist society" Antonio, 1981, p. 381), it is incumbent on others to democratize and decolonize these constructions (Fanon, 1961; Freire, 1970; Nandy, 1987; Nkrumah, 1964; Smith, 1999) and take a standpoint outside the relations of ruling (Smith, 2005). We need to be clear about whose standpoint we are taking and why, whose questions need to be addressed and what for and then write with responsibility toward those for whom we claim to write.

With this preamble and invocation in mind, the idea for this collection on *Education, Participatory Action Research, and Social Change: International Perspectives* was prompted by the critical observation that research, including its more participatory expressions, as it pertains to education and social research at large, is increasingly being defined (in concept and practice) by

(a) an onto-epistemic Euro-American cultural modernization imperative with its attendant homogenizing and assimilationist cultural-educational-research implications (Grande, 2004; Nandy, 2005; Smith, 1999);
(b) a neoliberal market fundamentalism that corrals and instrumentalizes research toward selecting and addressing research issues that are primarily of significance to the market and to corporatized-states (or conversely, research that is of little political threat to these interests) to assist in the management and control of populations (a sociology of regulation and discipline) (Baxi, 2000; Hamm, 2005; Levidow, 2005; Reimer, 2004; Sears, 2003) in service of global capital and modernization agendas, while continuing to proclaim the myths of democracy and sustainability;
(c) an increasing Euro-American professionalization (e.g., technicization and scientization of method) of the practice/approach to participatory forms of research (or approaches to an "acceptable PAR") that submerges the ideological content of theories and knowledge claims and masquerades for a purported *objectivity* (while simultaneously denouncing the possibility of "objective truths and methods") that is subsequently deemed funds-worthy (for examples of techno-managerial PAR designs, see selections from Reason & Bradbury, 2007), while the object and purpose of research is curtailed to servicing the professions (Kemmis & Smith, 2008) (e.g., enhancements in teacher pedagogy, health practice, shop-floor training programs, and "quality of work life" programs for productivity and effective service delivery) in the Lewinian tradition of action research, thereby effectively circumventing issues of power-political interests, the relations of power and how research contributes to the reproduction of sociopolitical asymmetries; and
(d) the co-optation of "participatory *anything*" (essentially processes of democratization) in to its antithesis, that is, processes of control and discipline, benign or exploitative, by international institutions allegedly addressing progressive concerns around Third World development, debt, impoverishment, and inequality (e.g., World Bank conceptions and applications) (Abrahamsen, 2000; Cooke & Kothari, 2001; Jordan, 2003; Rahnema, 1990) or by civil society groups (e.g., international NGOs) (Green, 2000; Kamat, 2002) peddling empowerment, justice, and human rights, while simultaneously working toward depoliticizing the politics of pain (tranquilizing protest with the rhetoric of the promise of participation) often engendered by market-state violence,[1] largely left unaddressed by the voluntary champions of the people or alternatively, activated as fodder for self-perpetuation and continued aid-dependent relationships.

The attempted political dismemberment and disengagement of PAR (especially in relation to anti/critical colonial variations) in professional and academic spaces is, both, predictable (given the politics of co-optation and its role in helping to reproduce a global capitalist hegemony and Euro-American knowledge hegemony) and ironic, given the real extent of marginalization, dispossession, and colonization (Bond, 2006; Chossudovsky, 2003; Rodney, 1982) in the contemporary era and more significantly, the growing and vocal resistance to attempted subjugations (Barker, 2005; Polet, 2007). Such spaces of a people's PAR and social action, points to the possibilities for academic versions of PAR to become engaged with the politics of the margins (Kane, 2000; Kapoor, 2008) as these sites remain the soil of germination for indigenous, anti/critical colonial, and third worldist PAR engagements.

The neutering of the political intent and definition of PAR (in academia and in governing institutions) processes committed to the interests of a popular pedagogical and knowledge project of, from, by, and for marginalized and dispossessed (in economic, political, cultural, ecological, and spiritual terms) social groups, calls attention to the pressing political and epistemic necessity to redraw the lines distinguishing various PAR engagements. This task needs to be undertaken with a view to revive and revitalize (in academic discourse) the notion and the contemporary reality that educational research and educational PAR can and is being utilized in the politico-cultural pedagogies and the daily praxis of social groups addressing the impositions of modernizing cultural imperatives of formal education and research (including increasingly professionalized forms of academic PAR, see Reason & Bradbury, 2006)[2] that continue to remain tainted by imperial and colonial ambition; an ambition tied to Eurocentric knowledge production and to the march of capital, euphemistically referred to as globalization (Hamm, 2005; Petras & Veltmeyer, 2001).

This critical assessment of contemporary versions of PAR and participatory forms of research in education and social research at large, prompted the development of this collection of chapters drawing primarily from indigenous, global South, and Euro-American critical traditions in educational and social research that attempt to (1) demonstrate what PAR, education, and social change mean in varied international contexts (while foregrounding indigenous and South contexts), since research germinates from different culturally defined and politico-historically located onto-epistemic starting points and political projects; (2) emphasize the political nature and contribution of socioeducational research and the different ways in which PAR can and is contributing to various political projects of the *wretched of the earth* (Fanon, 1961) internationally, while continuing to rely on a critical reflexivity (a defining ingredient of any PAR process committed to a politics of the margins and the constant reinvigoration of participatory research approaches) concerned with the moral, practical, and political ramifications and contradictions of PAR activations; (3) demonstrate and explore aspects of the critique of current approaches to research (including dominant conceptions of its more participatory variants) and the associated political interests that are served by such research; and (4) address these afore-mentioned possibilities while drawing upon participatory educational (formal, nonformal, informal, and incidental learning) and sociocultural research in multiple spaces including formal education (e.g., higher education), community (e.g., in indigenous communities) and applications in and from social movements/struggles, primarily in the South.

By embracing indigenous conceptions, approaches and practices of PAR as a living praxis; by magnifying the role and contribution of PAR in the multifarious struggles of marginalized social groups in the regions of the global South (Africa, Asia, and Latin America); and by engaging critical Euro-American conceptions of PAR and its utility in a politics attentive to addressing ecological concerns, commercialization of education/research and the containment of democratic pedagogies and popular research/knowledge processes in formal education, it is hoped that this collection will return PAR to its anti and/or critical-colonial roots in living indigenous traditions (Smith, 1999), to Euro-American critical traditions and to third worldist conceptions (Fals-Borda, 1979, 1981; Fals-Borda & Rahman, 1991; Freire, 1970; Hall, Gillette, & Tandon, 1982; Mustafa, 1981; Nyerere, 1979; Rahman, 1985) from which PAR germinated as a politics, a pedagogy and a knowledge of, by, and for the people.

Specifically, the collection attempts to address some of the following guiding questions/issues related to education, PAR, and social change in international and marginal contexts:

(i) What is PAR in contemporary international and marginal contexts? How is it being conceptualized in different education and social change locations and initiatives? What are some examples of these various PAR projects for social change?
(ii) Who are the protagonists of PAR work and in what communities of engagement? How does the practice of PAR in indigenous and Southern locations continue to make research and education/learning meaningful to the participating communities? What do indigenous and diasporic authors from the South have to say about PAR from their different locations of partnership?
(iii) From critical vantage points, what are some of the current preoccupations and issues concerning PAR and the politics of knowledge creation, education, and social change?

Organization of the Book

The book is divided into two sections, one on theory and the second on case studies drawn from Africa, Latin America, Asia, and Canada.

This division should not be taken too literally, however, as each of the seventeen chapters that comprise the book have in one way or another attempted to show that PAR is a composite methodology that is concerned foremost with questions of *praxis,* that is, the integration of theory and practice as an ongoing problem to be addressed in any form of social inquiry, whether in ethnographies and other kinds of fieldwork, or in understanding the generation of theory as a practical activity. In this sense the division is ultimately heuristic in simply allowing each author to bring a particular focus to their writing as an expression of a dimension of PAR. With this caveat in mind, we now proceed to provide a brief overview of the themes and issues that constitute the book.

Rather than present a chapter by chapter overview as is customary in many introductions, we have chosen to draw out and present what we view as the salient preoccupations and issues that are common to the authors who have made contributions to this book on PAR. In doing this, we hope to show that not only has PAR evolved and transformed considerably since its beginnings more than fifty years ago, but that in the contemporary period it continues to be the most effective form of research for working with marginalized communities and populations across a wide range of social/geographical contexts, and on an equally wide array of pressing issues from street kids in Brazil, rural women in Tanzania, indigenous populations in Algeria, Canada, India, and New Zealand, to communities affected by multinational mining operations in Ghana. Broadly speaking, these preoccupations and issues identified by authors fall into the following categories, each of which we discuss in the following text: co-optation and assimilation; knowledge creation and the critique of mainstream social sciences; social movement learning and PAR; and PAR as an indigenous methodology.

Co-optation and Assimilation

This is a theme that cuts across several of the chapters contained in the first section of the book on theory (e.g., Jordan, Kapoor, Kincheloe, Lange, McTaggart and Curró) and which is also explored in some of the case studies (e.g., Barua, Chovanec and Gonzalez, Veissiere, and Walsh). Broadly, the analysis is that since its inception almost half a century ago within anticolonial movements in the global south, PAR increasingly has been subject to forces that have compromised its revolutionary potential as a transformatory methodology for subaltern and otherwise marginalized populations aimed at bettering their social and political conditions. For example, it is clear from both Jordan's and Kapoor's analysis of, respectively the World Bank and the Adivasi ("untouchables") of the Indian state of Orissa, that there is a real and palpable tension in both the trajectory of PAR and its uses as a tool for particular kinds of social change. In the former instance, it is evident that cloaked within the apparently progressive academic discourses of social capital used by the World Bank (and many other international development agencies and NGOs), the concept and practice of participation, unequivocally, has been subordinated to a neoliberal agenda that in many respects mirrors the aims, objectives, and priorities of nineteenth- and twentieth-century colonialism. This kind of analysis is also explored in Chovanec and Gonzalez's study of young women activists in the northern Chilean city of Arica. They note that the young women activists that they worked with had become increasingly critical of the imperialist illusion of participation orchestrated by successive liberal-democratic governments installed after civil rule was reestablished in 1989. For these women, a critical praxis of action and reflection became central to developing a critique of neoliberal policies that continues unabated, despite the demise of the Pinochet dictatorship that occurred after this period. The analysis presented by Chovanec and Gonzalez is that in order to counter neoliberal constructions of participation generated by institutions such as the World Bank, practitioners of PAR must engage the groups and populations with whom they work in a systematic analysis of the social relations of capitalism to prevent cynicism, disillusionment, and disengagement. Kapoor's argument is that it is important to distinguish between what he terms as participatory academic research (or par) as opposed to PAR arising from an embedded and organic process rooted in the concerns of marginalized groups. For academics using participatory research methods with subaltern groups such as the Adivasi, this inevitably poses questions about in whose interests the research process is being organized. For Kapoor the only way to ensure that a PAR process is initiated and sustained is for academic researchers to continually work at embedding all aspects of participatory research in a living praxis, where participants learn to take control and academic researchers become willing hostages to their concerns. Other dimensions of this tension, between the rationalities and imperatives driving funded academic research and the concerns and issues of particular groups and populations who constitute the focus of participatory research, is also explored in the chapters by Kincheloe, Lange, Barua, Veissiere, and Walsh. Again, in each of these chapters the authors discuss the tendency for participatory forms of research to become embroiled in a politics of co-optation that is exercised either through the disciplinary character and history of their research paradigm (e.g., Veissiere and Walsh in relation to Anthropology), or the social organization of participation itself (e.g., Kincheloe, Lange, Barua). The point made by all these authors, however, is that in participatory research processes it is the politics of research that has to be actively engaged with to prevent co-option and subordination to the disciplinary knowledge making processes of Western scholarship.

Knowledge Creation and the Critique of Mainstream Social Sciences

A number of chapters in this collection focus on two interrelated themes that have historically defined PAR: (1) PAR's role in creating new forms of knowledge from the standpoint of subaltern groups; (2) its relationship to the mainstream social sciences. Weber-Pillwax, for example, provides a reflective account of her experiences in dealing with various levels of local and regional government, and universities, in trying to secure acceptance and legitimacy for indigenous (Cree) ways of knowing, viewing, and experiencing the social world within a predominantly monocultural educational system. As she observes in her account of the internal struggles she engaged in with Northland School Board over the 1980s and 1990s, the indigenous participatory elements of her culture provided the intellectual and spiritual resources to confront, challenge, and negotiate spaces for Cree knowledge with the dominant administrative apparatuses of the school system. In this respect, PAR for her, represents an everyday revolution that facilitates the movement of aboriginal people between their indigenous culture and the institutions of the colonial power. Chambers and Balanoff in their contribution make a similar argument in relation to their *Ulukhaktok Literacies Research Project*. The legacies of colonization by the Canadian state, they observe, has had profoundly negative implications for the Inuit people that they work and research with in Northern Canada. In particular, they note that the introduction of an educational system imposed from the South effectively disrupted and fragmented traditional knowledge producing practices that had historically been part of Inuit culture. It is against this background that they initiated their PAR-based *Ulukhaktok Literacies Research Project* as a means to reinvigorate traditional approaches to learning and knowledge creation that are both relevant and meaningful within the broader context of Inuit culture. They also point out, however, that the criteria used by Canadian federal and provincial funding agencies do not only mitigate against supporting PAR projects such as theirs, but also work to undermine indigenous intellectuals from leading them as Principal Investigators. In this respect the conceptual practices of the social sciences imposes a double deficit on research within indigenous communities. This is a view echoed by Conrad and Kendall in their piece on *Making Space for Youth*, where they note that not only it is conventional social scientific research (e.g., survey questionnaires) "problematic and inappropriate [...] yielding ineffectual results" on homeless/street involved youth in Canada, but that it is often utilized as a "stalling tactic" by "tight fisted" governments who are reluctant or averse to fund policies that constructively engage with the problem. As a consequence, Kendall and Conrad's response has been to establish PAR partnerships between iHuman (an NGO) and university researchers that engage homeless/street involved youth in arts-based, participatory projects, that draw on their oral traditions, stories, life histories, photographs, radio, music, myths, and so on, in making digital art video narratives of their lives. Barua, Belkacem, and Shizha's chapters deal with the issue of how indigenous cultures with strong participatory traditions—and their respective knowledge generating traditions, customs, and practices—collide and come into conflict with hegemonic systems of thought that either undermine or repress them. Barua discusses this process in relation to the changing character of funding and policy direction of NGOs concerning micro-credit in Bangladesh; for Belkacem it concerns the contradictory situation of the Berber, caught between fundamentalist Islam and state surveillance; whereas Shizha's study focuses on how Western conceptual practices of science education imposed by colonization have systematically excluded indigenous knowledges from the school curriculum, despite their obvious relevance to contemporary debates on sustainable development.

Last, it is important to draw attention to the interdisciplinary and experimental knowledge making practices that contributors have reported on in this book. These range from Walsh's use of video documentary techniques with shack dwellers in Durban (South Africa), arts-based techniques employed by Harata and Greenwood in developing Maori literacy programs for teachers in New Zealand, to the construction of digital life histories that Conrad and Kendall explored with homeless youth in Edmonton (Canada). What is striking about all these examples is that practitioners of PAR are experimenting with methodologies and methods that are far more innovative than those used by conventional social science or educational research.

Social Movement Learning and PAR

Perhaps the most influential work on adult learning in recent years is Foley's (1999) *Learning in Social Action*. Foley's insistence on the pivotal role of informal learning in the everyday lives of people under capitalism has had profound effects on both research conducted on the learning that adults do within particular contexts (work, leisure, domestic contexts) and the way in which learning as a complex, multifaceted activity, has been understood. Foley's observations that learning—particularly informal learning—is an integral aspect of all human activity and that it is shaped by interpersonal, institutional, and broader social and political forces (see chapter by Jordan), has clear parallels with what most, if not all, authors in this compilation have to say about PAR as an activity concerned with the generation of self-reflexive learning among its constituent groups and populations. In this respect, the book contributes, both in theory and through the case studies it presents, to the debate concerning the relationship of learning to research, politics, and social change within capitalist societies. There are numerous examples of this kind of analysis throughout the collection. Valerie Kwai Pun's study of WACAM (Wassa Association of Communities Affected by Mining), the NGO that spontaneously emerged in Ghana in 1998, reveals how a PAR inspired national popular education movement involving sixty communities and more than ten thousand people was created to contest and challenge the negative social and environmental impacts of gold mining by multinational corporations. As she convincingly shows, through its provision of workshops, training programs, and community sensitization projects, WACAM has congealed into a national social movement constituting a culture of resistance to the incursions of unbridled capitalist development within Ghana over the last decade. Similar kinds of community activated anticapitalist/ neoliberal learning are also analyzed by Chovanec and Gonzalez in their study of women's social movements in Chile, and by Kapoor in his study of indigenous Adivasi in Orissa. Another dimension of the learning that PAR can generate is provided by Conrad and Kendall's study of homeless/ street involved youth, and Lange's work in her university extension classes with middle-class professionals. In both instances, reflexive-participatory methods are used to bring self-awareness and heightened consciousness of the circumstances in which individuals find themselves, whether as homeless aboriginal youth on the streets of Edmonton, or as questioning middle-class adults participating in university extension courses focused on sustainability. What these and other chapters in this collection show is that (1) PAR and learning are inseparable activities that are embedded within a tight dialectical relationship of mutual change and transformation; (2) PAR can, under certain circumstances, become a powerful tool for the generation of critical and otherwise anticapitalist forms of learning for both individuals and communities; (3) informal learning is key to understanding the complex dimensions of knowledge creation within social movements.

Indigenous Peoples and PAR

As Smith (1999) has shown, while indigenous people have strong and ongoing traditions of direct participation within their cultures, these have largely disappeared within industrial capitalist societies with the advance of technical rationality, instrumental reason, and the proliferation of the wage labor-capital relation. As discussed by a number of authors in this collection, this process has been one of uneven development in the sense that while increasing numbers of aboriginal communities around the globe have become increasingly subject to penetration by these processes and relations, there are still others that have protected and maintained the integrity of their traditions, languages, and cultures. Consequently, some aboriginal communities have managed to maintain and reproduce social relations and practices that effectively constitute organic forms of PAR that are specific to the indigenous cultures that generate them. Contributions to this volume by Conrad and Kendall, Te Aika Harata and Greenwood, Chambers and Balanoff, Belkacem, Kapoor, Kwai Pun, Shizha, Mhina, and Weber-Pillwax show that these cultures of PAR do not only act as deep reservoirs in the reproduction of language, knowledge/skill, experience, and understanding, but they also provide local resources for resistance and negotiation against the globalizing tendencies of neoliberalism across a wide range of contexts in both the North and Global South. Among other things, what they appear to do most effectively is to provide strong cultural spaces that are compatible with research-based forms of PAR described by contributors to this volume. For example, Weber-Pillwax perceives strong continuities between Cree notions of communal space denoted by *sakaw*, as does Mhina and Shizha in their descriptions of, respectively, *Tweyambe* and *dare/indaba*. What is significant about these cultural spaces is that while they are constituted by social relations that express research-based approaches to PAR (e.g., direct participation, democratic discussion and debate, inclusiveness, a concern with social justice etc.), they also pose alternative sources of knowledge-producing practices from dominant Western sciences. Smith (1999) has already shown how such practices might be used to reconceptualize the ways in which research is conducted in relation to aboriginal peoples, as well as the dominant culture, by rejecting predominantly positivist and structuralist paradigms in favor of approaches that attempt to work outside and are critical of the hegemonic discourses of the academy. Such an indigenous research methodology (IRM) as Weber-Pillwax calls it, is not just simply a set of technical, how to, methods for conducting forms of PAR, but emphasizes the fundamental connections between history, culture, politics, and research, that is, that any kind of research is laden with questions of value, human judgment, and politics irrespective of its disciplinary grounding.

Notes

1. For example, according to the World Commission on Dams (reference), dams alone have displaced 80 million people reducing them to "development refugees" (Rajagopal, 2004).
2. As Rahman (1985) noted in the early 1980s, "PAR, after all, is threatening to become a respectable intellectual movement, and participatory researchers are gaining in social status, within and across national frontiers. PAR is getting institutionalized" (p. 19). These observations remain ever more pertinent today, given the neoliberalization of higher education, an associated culture of performativity and an academic entrepreneurialism that encourages co-optations and even within PAR circle flirtations with professionalized pseudo-scientific social research (e.g., Sage handbooks on research methods are achieving almost biblical status in the chase for research funds and research acceptability).

References

Abrahamsen, R. (2000). *Disciplining democracy: Development discourse and good governance in Africa*. London: Zed.

Antonio, R. (1981). Immanent critique as the core of critical theory: Its origins and development in Hegel, Marx, and contemporary thought. *British Journal of Sociology, 32*(3), 325–351.

Barker, J. (2005). *Sovereignty matters: Locations of contestation and possibility in indigenous struggles for self-determination*. Lincoln: University of Nebraska Press.

Baxi, U. (2000). Human rights: Suffering between the movements and markets. In R. Cohen, & S. Rai (Eds.), *Global social movements* (pp. 33–45). New Brunswick, NJ: Athlone Press.

Bond, P. (2006). *Looting Africa: The economics of exploitation*. London: Zed.

Chossudovsky, M. (2003). *The globalization of poverty: Impacts of IMF and World Bank reforms*. New Jersey: Zed.

Cooke, B., & Kothari, U. (Eds.). (2001). *Participation: The new tyranny?* London: Zed.

Fals-Borda, O. (1979). Investigating reality in order to transform it. *Dialectical Anthropology, 4*(1), 33–56.

———. (1981). The challenge of action research. *Development: Seeds of Change, 1*, 55–61.
Fals Borda, O., & Rahman, M. (Eds.). (1991). *Action and knowledge: Breaking the monopoly with participatory action-research*. New York: Apex Press.
Fanon, F. (1961). *Wretched of the earth*. New York: Grove Press.
Freire, P. (1970). *Pedagogy of the oppressed*. New York: Continuum.
Grande, S. (2004). *Red pedagogy: Native American social and political thought*. Lanham, MD: Rowman & Littlefield.
Green, M. (2000). Participatory development and the appropriation of agency in Southern Tanzania. *Critique of Anthropology, 20*(1), 67–89.
Hall, B., Gillette, A., & Tandon, R. (1982). *Creating knowledge: A monopoly?* New Delhi, India: Society for Participatory Research.
Hamm, B. (2005). Cynical science: Science and truth as cultural imperialism. In B. Hamm, & R. Smandych (Eds.), *Cultural imperialism: Essays on the political economy of cultural domination* (pp. 60–76). Toronto: Broadview Press.
Jordan, S. (2003). Who stole my methodology: Co-opting PAR. *Globalisation, Societies and Education, 1*(2), 185–200.
Kamat, S. (2002). *Development hegemony: NGOs and the state in India*. Delhi: Oxford.
Kane, L. (2000). Popular education and the landless people's movement in Brazil (MST). *Studies in the Education of Adults, 32*(1), 36–50.
Kapoor, D. (2008). Globalization, dispossession and subaltern social movement (SSM) learning in the South. In A. Abdi, & D. Kapoor (Eds.), *Global perspectives on adult education* (pp. 100–132). New York: Palgrave Macmillan.
Kemmis, S., & Smith, (Eds.). (2008). *Enabling praxis: Challenges for education*. Rotterdam: Sense Publications.
Levidow, R. (2005). Neoliberal agendas for higher education. In A. Saad-Filho, & D. Johnston (Eds.), *Neoliberalism: A critical reader* (pp. 156–162). London: Pluto.
Mustafa, K. (1981). *Participatory research among pastoralist peasants in Tanzania: The experience of the Jipemoyo project in Bagarnoyo district*. Mimeographed World Employment Programme Research Report, ILO, Geneva.
Nandy, A. (1987). *Traditions, tyranny and utopias: Essays in the politics of awareness*. New York: Oxford University Press.
———. (2005). Imperialism as a theory of the future. In B. Hamm, & R. Smandych (Eds.), *Cultural imperialism: Essays on the political economy of cultural domination* (pp. 52–59). Toronto: Broadview Press.
Nkrumah, K. (1964). *Consciencism: Philosophy and ideology for de-colonization*. New York: Monthly Review.
Nyerere J. K. (1979). Adult education and development. In H. Hinzen, & V. H. Hundsdorfer (Eds.), *The Tanzanian experience: Education for liberation and development*. UNESCO Institute for Education, Hamburg: Evans Brothers London.
Petras, J, & Veltmeyer, H. (2001). *Globalization unmasked: Imperialism in the 21st century*. Halifax, Canada: Fernwood Press.
Polet, F. (2007). *The state of resistance: Popular struggles in the global south*. London: Pluto.
Rahman, M. (1985). The theory and practice of participatory action research. In O. Fals-Borda (Ed.), *The challenge of social change*. London: Sage.
Rahnema, M. (1990). Participatory action research: The "last temptation of saint" development. *Alternatives, 15*(2), 199–226.
Rajagopal, B. (2004). *International law from below: Development, social movements and third world resistance*. Cambridge: Cambridge University Press.
Reason, P., & Bradbury, H. (2006). *Handbook of action research*. Thousand Oaks, CA: Sage.
———. (2007). *The Sage handbook of action research: Participation, inquiry and practice*. Thousand Oaks, CA: Sage.
Reimer, M. (Ed.). (2004). *Inside corporate U, Women in the academy speak out*. Toronto: Sumach.
Rodney, W. (1982). *How Europe underdeveloped Africa*. Washington, DC: Howard University Press.
Sears, A. (2003). *Re-tooling the mind factory: Education in a lean state*. Aurora, ON: Garamond.
Smith, L. (1999). *Decolonizing methodologies: Research and indigenous peoples*. London: Zed.
Smith, D. (2005). *Institutional ethnography: A sociology for people*. Oxford: AltaMira.

PART I

International Perspectives on Education and PAR

CHAPTER TWO

From a Methodology of the Margins to Neoliberal Appropriation and Beyond: The Lineages of PAR

STEVEN JORDAN

This chapter builds on theoretical research that I have conducted over the past decade on both critical ethnography and participatory action research (PAR) (Jordan, 2002, 2003b). Originally, the aim of this research was to problematize what I perceived to be taken-for-granted practices in the conduct of mainstream qualitative research that were still shaped and informed by colonial relations generated by nineteenth-century anthropology and ethnography (Jordan, 1993). I became increasingly aware, however, that these relations were not simply historical residues that were specific to the conceptual practices of anthropology, but were more widely distributed throughout the social sciences, as Smith's work in relation to sociology has amply demonstrated (Smith, 1987, 1990a, 1990b, 1994). As a consequence I turned my attention to forms of PAR that, historically, had been generated outside the academy by social movements in the Global South that appeared to hold out the possibility of eschewing the colonial legacy of mainstream social sciences.

Despite the fact that approaches to PAR were largely generated and defined through the work of radical groups working within anticolonial movements in the Global South from the 1960s, it was clear that by the 1990s their legacy increasingly had been subject to a subtle process of institutionalization and co-option by mainstream social science researchers, private consultants, government bodies, international development agencies, and nongovernmental organizations (NGOs). Depending on the context, this process expressed itself in different ways, but was most apparent in the gradual and insidious separation of PAR from its radical origins and depoliticization as a methodology of the margins. The general effect, however, was to assimilate and reconstitute approaches to PAR within conceptual practices and forms of social organization that were aligned with neoliberal globalization (Jordan, 2003b).

In light of these observations, the first section of the chapter will explore the defining themes and issues that have given PAR its contemporary character over the last half-century. Following this, the second section will explore the politics of co-optation of PAR under neoliberal globalization. Given that PAR historically has been susceptible to co-optation, the last section of the chapter will examine the possibilities for establishing a counterhegemonic methodology through a synthesis of PAR and other forms of critical research, such as critical ethnography (Jordan, 2003a) and learning in social action (Foley, 1999). The aim here will be to explore common grounds, as well as the tensions and challenges inherent in attempting a methodological synthesis, or hybridization, of PAR with these other methodologies.

Lineages

The origins and development of PAR (and other forms of participatory research) are both complex and difficult to map with any precision. This is not only because the term is used loosely and often interchangeably with concepts such as action research, but it is also due to the fact that PAR is itself a blend of a broad range of research approaches and epistemologies that include participatory research, action research, feminist praxis, critical ethnography, aboriginal research methodologies, transformative education, critical and eco-pedagogies, and popular and community education. However, it is also worth noting that versions of PAR and more generically, participatory research, have not always been allied with radical or liberatory political movements, nor have they necessarily been aimed at the emancipation of oppressed and marginalized groups. As I will argue in the following text, participatory research has increasingly been the subject of co-option and assimilation by what the Canadian sociologist, Dorothy Smith (1994), has referred to as the relations of ruling.

Despite this blend of theoretical traditions and emerging tensions in the way in which it has been used, it is nevertheless possible to outline some general contours and key features that have marked PAR's historical development over the last fifty years. First, it is clear that the impetus for exploring forms of participatory research—though they were not necessarily named as such—came from the third world in the early 1960s. Inspired by anticolonial and related political struggles, scholars such as Paulo Freire (1972) and Orlando Fals-Borda (1969) began to focus their attention on how social science research could be used to relocate the everyday experiences and struggles of the poor, oppressed, and marginalized from the periphery to the centre of social inquiry. Within this scenario, social research was to be transformed from an abstract, detached, disinterested, and objective science conducted by outside experts into an emancipatory process centered on what Freire called conscientisation, where the poor were to become agents of social and political transformation aimed at creating just, peaceful, and democratic societies.

Second, independence from colonial powers invariably led to the emergence of forms of popular education through national literacy campaigns, such as those led by Castro in Cuba and the Sandinistas in Nicaragua. The aim of these literacy campaigns was not merely to inculcate functional literacy in the populations of the Global South, but to foster forms of popular consciousness that were critical, emancipatory, and democratic. The general thrust of these movements, it should be emphasized, were not only radical but also revolutionary (i.e., they had as their focus societal transformation). These developments have had their counterpart in the countries of the Global North. While not driven by the anticolonial and revolutionary contexts of the South, forms of adult and community education, labor/union programs, transformative education, green and ecology movements, and more recently the emergence of an international network of indigenous and antiglobalization groups have informed the politics of PAR through their commitment to a communitarian ethics of organization and practice.

A third strand in the development of PAR relates to its action component. Although the history of action research is connected with the development of PAR, it nevertheless can be distinguished from it in three important ways. First, action research has primarily European and North American origins. Second, it has been principally developed by academic researchers working from universities within the advanced capitalist world of the Global North. Third, its ideological orientation has tended to be liberal, focusing on the improvement of professional practices—this is why it has proven to be so popular among researchers working with teachers and other professional groups. However, in recent years action researchers have also become concerned with issues of social justice that have shaped PAR. Although much action research continues to express its traditionally liberal/professional focus, there are a significant number of action researchers who have attempted to incorporate the radical lessons of both participatory research and popular education within their practice (Carr & Kemmis, 1993).

These themes—origins in the Global South, societal transformation, and alignment with action research—cannot capture all the complexities of the development of PAR over the last fifty years. They do serve to show, however, that it has been driven by a dynamic that has centered on a democratic, critical, and emancipatory impulse quite distinct from conventional research methodologies in the social sciences. Despite this, my argument in the next section will be that in the era of neoliberal globalization commencing in the late 1970s, approaches to participatory research, including PAR, have increasingly been subject to social, economic, and political forces that have either challenged or systematically compromised this legacy. It is to this process of co-option and assimilation that I now turn.

The Neoliberal Appropriation of PAR

Despite its social origins and radical traditions, PAR and other forms of participatory research have increasingly been subject to a subtle process of institutionalization and co-option by mainstream social science researchers, private consultants, government bodies, international development agencies, and NGOs. Depending on the context, this process has expressed itself in different ways. However, the general effect has been to assimilate and reconstitute these methodologies within conceptual practices and forms of social organization that articulate with the relations of ruling. This tendency is most clearly expressed in industrial sociology where PAR, or variations of it, has been used to explore the effects of shop-floor workers' participation in managerial decision making through the 1980s and 1990s. For example, Whyte's (1991) rendering of PAR, or that of Argyris' and Schön's Action Science (1991), reveal an overt concern with organizational change and learning on terms and conditions established by multinational corporations (e.g., Xerox). In this context, PAR's unique contribution, as part of the tool kit of the social sciences, is to co-opt workers' knowledge and understanding of the labor process to effect paradigm shifts on how to boost productivity and competitiveness (Whyte, 1991). Despite the human relations approach of these authors, the overall impact of the research process that they elaborate is to reconstitute PAR as a tool of capitalist accumulation.

The approach to participatory research outlined earlier, it should be noted, has inspired an array of participatory workplace initiatives that have attempted to reorganize and mask the historical tensions implicit in the capital-labor relation in the contemporary period. In particular, I am thinking of Senge's (1990) development of the concept of the learning organization and its widespread application to understanding and analyzing capitalist restructuring of the workplace. However, as Fenwick has pointed out, not only is it largely uncritical of this process, "the learning organisation concept emphasises productivity, efficiency and competitive advantage at the expense of the worker" (Fenwick, 1998, p. 151). Another, perhaps more insidious, example of what I am referring to is the rise of the team concept and the assumed benefits that it provides workers through participatory management. Yet as studies, such as Rinehart, Huxley, and Robertson (1997) have shown, the team concept invariably implies intensification of the labor process, greater stress through multitasking, and burn-out among workers. It is in these ways that the concepts of the learning organization and team concept can be considered as constitutive of the conceptual practices of power (Smith, 1990b).

As I have suggested earlier, while this process is not new it does appear to be both accelerating and proliferating (Davis & Monk, 2007; Harvey, 2005; Howard & King, 2008). Although it is not altogether clear why this has occurred in the contemporary period, it is evident that the gradual co-option of PAR has been coeval with the emergence and consolidation of neoliberalism as a hegemonic form of governance since the 1980s.

The rise of neoliberalism from the 1980s has had a profound impact on economics, politics, and culture within the advanced capitalist countries. Its proliferation throughout the 1990s in the former communist bloc, as well as less developed countries, now underpins the processes that constitute globalization. While neoliberalism can be viewed as a primarily economic philosophy in which free markets are the center-piece (Friedman & Friedman, 1985; Hayek, 1944), implicit within it is also a system of governmentality whose locus is the individual consumer, not the citizen of postwar social democracy. As theorists such as Yeatman (1997, 1998), Rose (1992), and Gamble (1986) have shown, the transition from the politics of citizen rights to one where everyday life is organized through consumers and market relations has fundamentally recontextualized the discourse of participatory democracy. By extension, my argument is that the prevalent discourses of participation that define contemporary approaches to PAR and participatory research have been increasingly infiltrated and appropriated by neoliberal discourses that have profoundly reconfigured the social relations of participation in the contemporary period.

Foucault (1991) has shown that regimes of governmentality are both insidious and subtle in the myriad of ways in which they generate and mould individual subjectivity and consciousness. In this respect neoliberalism over the last two decades has utilized a complex of ideological tools for this purpose. The most prominent and pervasive is that of the market which, as the historian E. P. Thompson (1991) has noted, is projected as

> an energising spirit—of differentiation, social mobility, individualisation, innovation, growth, freedom—like a kind of postal sorting-station with magical magnifying powers, which transforms each letter into a package and each package into a parcel. (p. 305)

To participate in a market is, therefore, to become subject to its "magical magnifying powers" through which individual consumers can realize their own subjectivities. In this way, the market has proven to be a powerful ideology in mobilizing consent for the implementation of a neoliberal agenda across a broad spectrum of economic, social, and educational policies that have commodified everyday life within global capitalism.

Although the market metaphor has constituted the master narrative, neoliberalism has also generated other conceptual practices that have saturated the social fabric. Of relevance to the discussion here are the concepts of social cohesion and social capital. The latter concept, in particular over recent years has become the darling of public policy debate across a wide range of contexts, particularly in relation to the idea of a learning society (Schuller & Field, 1998). Originally developed by Coleman (2000) in the 1960s, social capital was rediscovered by political scientists and sociologists in the 1990s (Portes, 2000). Since then, it has been embraced by a much broader spectrum of the academic community, governments, international organizations, and elements of civil society. While there are radical and conservative revisions of the concept, it nevertheless has been co-opted and assimilated within a neoliberal policy regime.

Recontextualized in this way, the idea of social capital has become a conceptual practice for legitimizing appropriate responses to the antisocial effects of the market and, therefore, reenergizing forms of participation within civil society from the standpoint of capital (Cooke & Kothari, 2001; Jordan, 2003b; Jordan & Yeomans, 1991). The World Bank's Web site pages devoted to social capital, for instance, provide a clear example of how the concept has come to be used to underpin forms of participatory development in its operations around the world. Indeed, its Social Capital Implementation Framework (SCIF) is quite specific in elaborating a development process that apparently is indebted to an ethic of popular participation for the countries it works in. In explicating its approach to Social Capital Measurement in Community-Driven Development (CDD) Operations for instance, it lays out the following process:

> When mapping social capital within a country, it is useful to look at the primary dimensions of social capital in the country and the existence of significant regional variations in how social capital is expressed across the country (such variations can have important operational implications). In addition to mapping of social capital, a practitioner identifying a potential project should analyze context-specific social capital for the area in which one plans to operate: Are there characteristics of pre-existing social capital that the project might usefully tap into that facilitate successful implementation? (World Bank, 2008)

Ostensibly this statement of intent appears both sensible and reasonable. As part of a broader methodological framework for measuring social capital in a region or country, it adopts a discursive framework that would not be unfamiliar to practitioners of PAR in its commitment to empowerment, promoting a culture of participation among the targeted beneficiaries, or the participation of marginalized groups; lessening social exclusion; increasing the capacity for collective action, and so on. Indeed, enhancing participation of formerly marginalized groups is the *raison d'être* that underpins how the World Bank construes and operationalizes the concept of social capital. However, the issue is not that the World Bank is committed to participation in its mapping of social capital, but who is being empowered to participate and under what conditions. As I have noted earlier (Jordan, 2003b), if these questions are asked, then it becomes evident that the deployment of the conceptual practices of social capital (e.g., participation, social inclusion, empowerment, and ownership) by the World Bank and related organizations such as the Asian Development Bank, the OECD (Organisation for Economic Co-operation and Development), or a wide range of NGOs, are an expression of a neoliberal world view within which the free-market, privatization, deregulation, and unfettered capital accumulation are projected as the only viable options for development and (post)modernization. Such analysis must also lead to the conclusion that the way in which these institutions construe civil society as somehow oppositional, or ameliorative of the worst impacts of neoliberal policies, has to be questioned.

Asking these and other questions of what appear to be seemingly benign and apparently progressive discourses can, therefore, lead us to a very different analytical standpoint. For example, the quote I have used earlier from the World Bank on mapping social capital within a country could be read as a modern version of what anthropology attempted to accomplish as an adjunct of colonialism in the nineteenth and first half of the twentieth centuries. As Kabbani (1986) puts it,

> Although Anthropology came to be a leveler of race and culture...it was inextricably linked to the functioning's of empire. Indeed, there can be no dispute that it emerged as a distinctive discipline at the beginning of the colonial era, that it became a flourishing academic profession towards its close, and that throughout its history its efforts were chiefly devoted to a description and analysis of non-European societies dominated by the West. It was the colonial cataloguing of goods; the anchoring of imperial possessions into discourse. (p. 62)

Thus, mapping social capital, its regional variations in how it is expressed across a country, as well as harnessing preexisting social capital is redolent of what Kabbani (1986) and others (e.g., Said, 1993) have described as the legacies of imperial and colonial rule. As Sadhvi and Cooke (2008) point out in relation to participatory management, the ways in which this is achieved in the contemporary period may have a different appearance, but the actual conceptual practices and social relations that catalogue and anchor these discourses have not. Their claim is that contrary to accepted orthodoxy, participatory processes used to give voice (this is also a theme on the World Bank Web page for social capital) to the poor and dispossessed through, for example, participatory management, are contiguous with forms of indirect rule exerted by colonial powers in the nineteenth and twentieth centuries. Indeed, the deployment of the discourses of participation, social capital, social inclusion, and civil society by the World Bank and other development institutions are to be understood as neocolonial in the way in which they attempt to map, reconfigure, and bind local customs, traditions, social relations, and practices through indirect forms of local rule exerted by neoliberal policies. The use of participatory methods to effect change within this policy context, as they point out, is not new. Indeed as Cooke (2008) in his chapter on participatory management demonstrates, a close examination of the work of Lewin and others (e.g., Collier) involved in the generation of action research and participatory methods in the immediate postwar period, reveals an overriding concern with power. This leads him to the conclusion that "[f]or all the claims of participation and empowerment, action research was still a means of controlling what the colonised did, according to the priorities of the colonial power" (p. 127). It would seem, therefore, that the origins of participatory methods (at least in the United States) were already implicated in technologies of power or what Corrigan has referred to as moral regulation (Corrigan, 1977; Corrigan & Corrigan, 1979; Corrigan & Sayer, 1985).

The argument presented so far, therefore, is that the neoliberal discourse of participation—organized through the conceptual practices of the market, social capital, and participatory development—continues to appropriate and reconstitute PAR in ways that are antithetical to its expressed principles, inclusive practices, and values. How this process may be either reversed, or at least ameliorated, is the topic of the next section.

Reclaiming PAR

Despite its history of marginalization within the Western social sciences, PAR and other forms of participatory research (e.g., community action research), have in recent years become the focus of increased attention from a wide range of government bodies, international development agencies, NGOs, and private management consultants. However, it is important to understand that the adoption of the concept and practice of participation has been radically recontextualized within the discourses employed by these organizations. As I noted earlier, we need only to consult the World Bank's Web site to understand how the discursive practices of participation (e.g., implicated in social capital) are now being used to exert forms of neoliberal governmentality through quasi-methodologies that resemble PAR. It is in this respect that PAR, and other forms of research that employ a participatory process (e.g., participatory management techniques), have been co-opted and reconstituted so that they are distinctly at odds with the emancipatory, indeed revolutionary, principles on which PAR was originally founded in the Global South. However, in this context it is also important to remember Cooke's (2008) observation on the origins of action research; that participatory methods were, from their very inception, entangled in questions of power and control.

If neoliberalism continues to drive forms of participation as it has done over the last three decades, then those of us committed to pursuing critical approaches to educational research need to reflect on what kinds of methodology are best suited for constructing PAR. As I indicated in the introduction to this chapter, my argument is that an approach derived from other methodologies, notably critical ethnography and Foley's (1999) notion of learning in social action are particularly well-matched for generating the methodological scaffolding for PAR that will resist the kinds of co-option and assimilation I have alluded to above. In what follows I outline and discuss the contours of what I am suggesting.

First, practitioners of PAR should explore and draw on other critical methodologies for the purposes of forming alliances with their respective research communities. One path that might be followed here is that of critical ethnography. Not only has the relatively short history of critical ethnography paralleled that of PAR's development, but its methodology, as well as the themes and issues it has focused on, are remarkably similar. For example, critical ethnography has mostly rejected positivism as a methodological approach to social research (although there are notable exceptions to this as I show in the following text), as it has notions of value-neutrality or objectivity. It also shares with PAR a strong ethical commitment to social justice. Consequently, critical ethnographers have found themselves working with the poor, marginalized, and otherwise subordinate groups within society. While it shares this common ground, critical ethnography has at least two important contributions that it can make to the development of PAR. The first is its integration of political economy within its theoretical and analytical framework. In an era of globalized capitalism, this is crucial to the analysis of social phenomena. Second, critical ethnographers have now developed an elaborate and sophisticated critique of the mainstream social sciences that shows how they are implicated in the relations of ruling under capitalism (Smith, 1990a, 1990b, 1994). This critique has increasingly been incorporated within the practices of PAR, providing it with a powerful rationale for its apparently nonconformist approach to social inquiry.

It also needs to be emphasized, however, that critical ethnography differs from PAR in several important ways, and that these differences may attenuate possibilities for such an alliance. As I have discussed the limitations of critical ethnography elsewhere (Jordan, 1993, 1996, 2002, 2003a), I will outline these only briefly in the following text. The first is that while critical ethnography shares a marginal status with PAR *vis a vis* the mainstream social sciences, its practitioners tend to be university trained academics concerned with the pursuit of research practices that conform to the traditions of scholarship as understood within the academy. One effect of this has rendered forms of critical ethnography that are imbued with positivism, while another has been to maintain and reproduce the division between expert/professional researcher and his/her subjects. This poses a second limitation. Because of its social location within universities and colleges—a position that increasing numbers of PAR researchers share should be noted—critical ethnography is often subject to and enmeshed within the hierarchical and highly individualized social relations of competitive research funding. This type of social organization obviously contradicts the democratic and communitarian ethos of PAR. A third limitation concerns the contemporary trajectory and orientation of critical ethnography, which has increasingly been defined by postmodernism. While this has not necessarily been a retrograde step, it nevertheless has accentuated highly individualized approaches to research (e.g., autoethnography) as well as a reification of theory over practice. This development raises difficult questions about the accessibility of critical ethnography to groups who do not possess the specialized training of university-based researchers. Despite these limitations, however, practitioners of PAR may still derive highly relevant insights from the study of critical ethnography for their own practice.

A second field of social inquiry that practitioners of PAR might pursue is marked out by learning in social action (Foley, 1999). For Foley, learning in social action has a several key elements. First, learning is viewed as an integral aspect of all forms of human activity. Second, a central feature of all human activity is informal learning, which is embedded in the routine practices of people in their everyday lives. Third, most learning emerges from struggles, which are shaped by interpersonal, institutional, and broader social and political forces. And last, learning to resist and struggle for better and just societies is possible, but this is inherently complex, contradictory, and contested. Unlike critical ethnography, therefore, which has emerged from disciplinary formations within the academy (e.g., anthropology), learning in social action is derived from a different social matrix composed of activists, the labor movement, and struggles of ordinary people over housing, health care, and the environment. Foley's focus on informal learning across these contexts is particularly relevant to PAR given what I have said earlier about the co-option of the discourse of participation by neoliberalism. That is, learning in social action urges us to pay closer attention to how neoliberal policies colonize local customs, traditions, practices, and social relations by harnessing them to forms of extra-local ruling. In particular, it poses questions about how and in what ways informal learning has been co-opted and made to work in the interests of social capital or the market as capitalist process, as well as generate resistance to it. Learning how to participate in the everyday world, and how this has been shaped by neoliberalism then becomes the focus of attention.

Last, practitioners of PAR need to pay close attention to the language and conceptual frameworks that they employ to develop their methodology. I have already pointed, for example, to the ambivalent legacy of the concept of social capital and its effects on the conceptual practices of participatory research. My argument is that we must be cognizant of the subtle effects that the use of mainstream concepts has on the research questions we pose, the research process we construct, and our analysis of the contexts that we explore. Such concepts and conceptual frameworks have, as I have shown, a profound influence on the processes of participatory research and the social organization of participation.

In the light of its complex and often contradictory history of co-option and assimilation, contemporary practitioners of participatory research would do well to acknowledge the critique of PAR that critical ethnographers have made—that power relations between professional researchers and participants are not necessarily equalized or erased through the mere act of participation. Thus, researchers need to approach studies that claim a participatory methodology cautiously and not assume that just because participatory methods have been used, that this necessarily endows it with an emancipatory impulse. As noted earlier, while the fact of participation in a research project may be compatible with Western values of empowerment, liberation, and democracy, it may also be equally bound to technologies of normalization, subjugation, control, and exploitation.

Conclusion

This chapter has, therefore, been concerned with three interrelated issues. First, it set out to provide a brief overview of the central themes and issues that have defined PAR over the past half-century. From a combination of popular anticolonial struggles, popular education, and action research, PAR emerged as a methodology not only of the margins, but also of the marginalized. As I showed, it was this legacy that endowed it with a mostly nonpositivist, critical, and overtly political character. Second, I showed how this critical tradition within PAR, and more generally participatory research, has increasingly become eclipsed by the emergence and pervasive influence of neoliberalism. Specifically, my argument was that the discourse of participation within capitalist democracies has been appropriated and recontextualized by neoliberalism, which in turn has had profoundly negative effects on the possibilities for participatory research in both the developed and less developed countries. Indeed, the forms of extra-local ruling exercised through the World Bank through the conceptual practices of social capital, for example, can be viewed as a contemporary form of neocolonialism. The third and last section looked at ways for reclaiming the critical traditions that originally defined PAR. My argument was that for PAR to become resistant to the discourses of neoliberalism, it has to critically engage with and incorporate the theory, methodologies, and methods of critical ethnography and learning in social action. In my view, unless practitioners of PAR are prepared to systematically engage in this kind of discussion and debate, it is likely that they will succumb to the processes of co-option and assimilation I have described in this chapter.

References

Argyris, C., & Schön, D. (1991). Participatory action research and action science compared: A commentary. In W. F. Whyte (Ed.), *Participatory action research* (pp. 85–96). Newbury Park: Sage.
Carr, W., & Kemmis, S. (1993). Action research in education. In M. Hammersley (Ed.), *Controversies in classroom research* (2nd ed., pp. 235–245). Buckingham: Open University Press.
Coleman, J. S. (2000). Social capital in the creation of human capital. In E. L. Lesser (Ed.), *Knowledge and social capital: Foundations and applications* (pp. 17–42). Boston: Butterworth Heinemann.
Cooke, B. (2008). Participatory management as colonial administration. In D. Sadhvi, & B. Cooke (Eds.), *The new development management* (pp. 111–128). London: Zed.
Cooke, B., & Kothari, U. (2001). The case for participation as tyranny. In B. Cooke, & U. Kothari (Eds.), *Participation: The new tyranny* (pp. 1–15). London: Zed.

Corrigan, P. (1977). *State formation and moral regulation in nineteenth-century Britain: Sociological investigations*. Unpublished Doctoral Dissertation, Durham University, Durham, England.

Corrigan, P., & Corrigan, V. (1979). *State formation and social policy until 1871*. London: Edward Arnold.

Corrigan, P., & Sayer, D. (1985). *The great arch*. London: Basil Blackwell.

Davis, M., & Monk, D. B. (2007). *Evil paradises: Dreamworlds of neoliberalism*. New York: New Press: Distributed by W. W. Norton & Co.

Fals-Borda, O. (1969). *Subversion and social change in Colombia* (J. D. Skiles, Trans.). New York: Columbia University Press.

Fenwick, T. (1998). Questioning the concept of the learning organisation. In S. M. Scott, B. Spencer, & A. M. Thomas (Eds.), *Learning for life: Canadian readings in adult education* (pp. 140–152). Toronto: Thompson Educational Publishing.

Foley, G. (1999). *Learning in social action: A contribution to understanding informal education*. London: Zed.

Foucault, M. (1991). Governmentality. In G. Burchell, C. Gordon, & P. Millaer (Eds.), *The Foucault effect: Studies in governmentality* (pp. 87–104). London: Harvester-Wheatsheaf.

Freire, P. (1972). *Pedagogy of the oppressed*. Harmondsworth: Penguin.

Friedman, M., & Friedman, R. (1985). *The tyranny of the status quo*. Harmondsworth: Penguin.

Gamble, A. (1986). The political economy of freedom. In R. Levitas (Ed.), *The ideology of the new right* (pp. 25–54). Oxford: Polity Press.

Harvey, D. (2005). *A brief history of neoliberalism*. Oxford: Oxford University Press.

Hayek, F. (1944). *The road to serfdom*. London: Routledge Kegan Paul.

Howard, M. C., & King, J. E. (2008). *The rise of neoliberalism in advanced capitalist economies: A materialist analysis*. Basingstoke, England; New York: Palgrave Macmillan.

Jordan, S. (1993, May). *Critical ethnography: Problems in contemporary theory and practice*. Paper presented at the Canadian Learneds, Carleton University (Ottawa).

———. (1996). *Schooling the vocational: The Technical Vocational Education Initiative (TVEI) and the making of the enterprise culture*. Unpublished Doctoral Disseration, McGill University, Montreal, Canada.

———. (2002). Critical ethnography and the sociology of education. In A. Antikainen, & C. Torres (Eds.), *Sociology of education: Perspectives for the new century* (pp. 82–100). Boulder: Rowman & Littlefield.

———. (2003a). Critical ethnography and educational research: Re-envisioning the sociology of education in an era of globalisation. *Education and Society, 21*(1), 25–52.

———. (2003b). Who stole my methodology: Co-opting PAR. *Globalisation, Societies and Education, 1*(2), 185–200.

Jordan, S., & Yeomans, D. (1991). Whither independent learning? The politics of curricular and pedagogical change in a polytechnic department. *Studies in Higher Education, 16*(3), 291–308.

Kabbani, R. (1986). *Europe's myths of orient*. Bloomington: Indiana University Press.

Portes, A. (2000). Social capital: Its origins and applications in modern sociology. In E. L. Lesser (Ed.), *Knowledge and social capital: Foundations and applications* (pp. 43–67). Boston: Butterworth Heinemann.

Rinehart, J., Huxley, C., & Robertson, D. (1997). *Just another car factory? Lean production and its discontents*. Ithaca, NY: Cornell University Press.

Rose, N. (1992). Governing the enterprising self. In P. Heelas, & P. Morrise (Eds.), *The values of the enterprise culture: The moral debate* (pp. 141–164). London: Routledge.

Sadhvi, D., & Cooke, B. (Eds.). (2008). *The new development management*. London: Zed Books.

Said, E. (1993). *Culture and imperialism*. New York: Knopf.

Schuller, T., & Field, J. (1998). Social capital, human capital and the learning society. *International Journal of Lifelong Education, 17*(4), 226–235.

Senge, P. (1990). *The fifth discipline: The art and practice of the learning organisation*. New York: Doubleday.

Smith, D. E. (1987). *The everyday as problematic: A feminist sociology*. Toronto: University of Toronto Press.

———. (1990a). *Texts, facts, and femininity: Exploring the relations of ruling*. London: Routledge.

———. (1990b). *The conceptual practices of power: A feminist sociology of knowledge.* Toronto: University of Toronto Press.

———. (1994). *The relations of ruling: A feminist inquiry.* Paper presented at the Group for Research into Institutionalization and Professionalization of Knowledge-Production (GRIP), University of Minnesota.

Thompson, E. P. (1991). *Customs in common.* London: Penguin.

Whyte, W. F. (Ed.). (1991). *Participatory action research.* Newbury Park: Sage.

World Bank. (2008). Social Capital Measurement in CDD Operations. Retrieved from http://web.worldbank.org/WBSITE/EXTERNAL/TOPICS/EXTSOCIALDEVELOPMENT/EXTTSOCIALCAPITAL/0,,contentMDK:20193038~isCURL:Y~menuPK:418220~pagePK:148956~piPK:216618~theSitePK:401015,00.html

Yeatman, A. (1997). Contract, status and personhood. In G. Davis, B. Sullivan, & A. Yeatman (Eds.), *The new contractualism?* (pp. 39–56). Melbourne: Macmillan.

———. (1998). Interpreting contemporary contractualism. In M. Dean, & B. Hindess (Eds.), *Governing Australia: Studies in the contemporary rationalities of government* (pp. 227–241). Melbourne: Cambridge University Press.

CHAPTER THREE

Participatory academic research (par) and People's Participatory Action Research (PAR): Research, Politicization, and Subaltern Social Movements in India

Dip Kapoor

Introduction

The Adivasi-Dalit Ekta Abhijan (ADEA) is a subaltern[1] social movement organization of some twenty-one thousand Kondh Adivasis (original-dwellers) and Panos (scheduled caste group, pejoratively referred to as untouchables as per caste-relegation as impure peoples) people in the southern region of the eastcoast state of Orissa, India. As a movement organization committed to the political activation of predominantly Kondh Adivasi and Panos communities located in more than 120 villages, the ADEA is a contemporary example of popular trans/local activism partially aimed at subaltern dispossession and displacement (material and cultural) by national development and neoliberal globalization (Kapoor, 2004, 2008, 2009) or modernizing socioeconomic processes predicated on the attempted dismemberment of subaltern communities,[2] albeit allegedly in their best interests and the greater public good.

> Even today you will find there is not enough cultivable land available for our people because they have taken it away...They have the power of *dhana* (wealth) and *astro-shastra* (armaments). They have the power of *kruthrima ain* (artificial laws and rules)—they created these laws just to maintain their own interests...and where we live, they call this area *adhusith* (or Adivasi infested, pejoratively understood as pest-infestation)...we are condemned to the life of *ananta paapi* (eternal sinners), as *colonkitha* (dirty/black/stained), as *ghruniya* (despised and hated). Adivasi leader of the ADEA (focus group discussion notes, 2007)

This chapter explores the connections between research and politicization in social movement contexts, elaborating on emergent political possibilities around funded participatory academic research (*par*) on the one hand or research which purports to understand participant constructions of socioeducational phenomena in partnership with them (participatory case study research in to "Learning in Adivasi social movements")[3] emerging from academic formulations designed to meet funding agency/university requirements and a people's PAR,[4] which moves to politicize politically sanitized *par* (the participatory case study in this instance) given the growing maturity of movement leadership in negotiating the political interests of their communities in the face of tremendous odds, as political-economic and cultural-spiritual (including research) encroachments remain a nagging part of Adivasi-Dalit (meaning downtrodden castes like the Panos) lives.[5] The proposition being advanced here is that when funded *par* is embedded in the living praxis of politically seasoned subaltern social movements given its (by definition) participatory intent, the movement participants can and will move to politicize the research process and engage with the research(er) in terms of a people's PAR informed by a subaltern politics, that is, the knowledge, learning, and collective action agenda of the ADEA steers funded *par* toward addressing the teleology of the movement. After all,

> Who are better prepared than the oppressed to understand the terrible significance of an oppressive society? Who suffer the effects of oppression more than the oppressed? Who can better understand the necessity for liberation? They will not gain this liberation by chance but through the praxis of their quest for it, through their recognition of the necessity to fight for it. (Freire, 1970, p. 29)

Specifically, this chapter shares some observations pertaining to an ADEA people's PAR and an engagement with a participatory academic research (*par*) process in relation to the following foci: (1) research origins, research processes, and movement interests; (2) research and knowledge-learning-action mobilizations; and (3) lived-theoretical constructions and theoretical hearing impairments. It is suggested that *par* can learn from and will amplify movement prospects, despite academic institutional (e.g., funding agent) requirements and ethical protocols, the leaching of political-intent/open partisanship in research and the general curbing of researcher enthusiasm for acts of solidarity with marginalized social groups because any form of *par* (even those that are well within the boundaries of academic research stipulations) with movements will, by definition, need to be partially embedded in a movement context. This arrangement creates opportunities for a politically mature group of movement actors to make the research process count for the movement before, during, and long after data collection/analysis is over.

Participatory academic research (*par*) and People's PAR: Research and Movement Politicization

The author's relationship with the Kondh Adivasi and Panos (Dalit) communities goes back to the early 1990s. The tangible development of an organized relationship did not emerge until 1996, when an organized partnership was attempted between subaltern groups in the current movement locale and a Canadian voluntary organization. The partnership was a modest attempt on the part of all participants to begin to organize today's maturing movement constituency with the long-term intention of establishing a popular activism for local and translocal autonomy, subaltern political articulation/assertion and a strong movement organization or the ADEA, which eventually emerged in the late 1990s to address subaltern unity and the need for a combined show of strength to persevere against historic and current colonial intrusions in the name of national development and neoliberal globalization. Initiatives undertaken by the partnership over the years have been in relation to organization, political-legal (extra-legal) assertions over land, forest, and water and the continued process of building subaltern unity in the face of constant challenges by the developmentalist-state (Kapoor, 2004), the tentacles of the market (Kapoor, 2009), and a caste-ethnicity-religious politics of division and racism aimed at Kondh-Panos solidarity (Kapoor, 2007). Formal and funded research engagements have never played a part as a movement vector, although a people's PAR and activations of people's knowledge, learning/pedagogy, and organized social action to address social-structural domination, has always been a part of the living movement process. Initial consideration around the possibilities of engaging in a conscious, formalized, institutional research process, with the possibility of funding attached was shared with the ADEA leadership in 2004—the author was then a recent tenure track faculty member facing institutional expectations to "do" research.

Research Origins, Research Processes, and Movement Interests: People's PAR Contests the Academic Limits of *par*

Once it became clear to the ADEA leadership that a formal *par* initiative would likely mean being studied for academic interests linked to the advancement of socioeducational knowledge in distant institutions (universities and research funding agencies abroad), albeit with their help in analyzing their own situation (participation?) and that the primary impetus for such a process of knowledge generation was centered outside their sociocultural and political space, there was much debate about the need for such an alien and impersonal exercise (people commented on the disconnectedness of such research and questioned why they should be studied) in the first place, with (participatory) or without their help.

Although perceived as an outsider with a healthy dose of insider status as well, given our tangible relationship for little over a decade, it became incumbent on me to spell out the likely boundaries in relation to funded academic research and democratic possibility in such knowledge engagements, including the role of ethical protocols that were geared toward the protection of the interests/safety of research participants based on a class sensibility and context far-removed from Adivasi-Dalit situations. The latter would mean, for instance, doing research in a manner that did not incite or instigate political action that would likely jeopardize or even harm vulnerable research participants. As one leader remarked on hearing this, "if they are so concerned about our well-being, perhaps they should stop studying us for their books and work with us to address our problems." This line of questioning was eventually taken up in relation to the possibility of a people's research institution that would serve the knowledge and active interests of the ADEA region and the movement. Subsequently, the Center for Research and Development Solidarity (CRDS) was inaugurated as an Adivasi-Dalit organization in 2006, committed to people's knowledge, education, and social action in keeping with ADEA movement aspirations. Still a fledgling affair, there is no lack of enthusiasm around the proposition and this became a key emphasis on the road toward ADEA interest around the author's proposed *par* in to "Learning in Adivasi movements," the proposal to be developed for potential research funding in Canada. CRDS is currently actively engaged in documenting caste atrocities against Dalits/Panos and instigating legal action (under the Scheduled Caste and Scheduled Tribes Atrocities Act) against perpetrators.

Similarly, discussions around the lines of inquiry pertaining to the study of movement learning were seized as opportunities to pursue the practical, applied, and reflective interests of the movement constituency, as academic questions of theoretical import and genesis (e.g., why questions around movement learning with little apparent applied or movement generative import) were politely brushed aside with silence, as the ADEA leadership emphasized "how" questions and reflective possibilities emerging from their own conceptions embedded in continuous acts of integration and synthesis as opposed to academic research reflections/inquiry prompted by partial/segmented (or delimited) searches for partial understandings pertaining to socioeducational phenomena. The search for knowledge by university-trained academics and related onto-epistemic persuasions (*par*) were continually reframed by a people's PAR into opportunities to document and expose Adivasi-Dalit issues in relation to social-structural domination and generally moved to ask how the *par* process being defined by the author (in relation to academic knowledge sources and criteria) could be of more utility and sense to the movement. To put it in the words of one Panos leader, "Since the university wishes to learn more about us, our ways and our situation, respectfully, perhaps the best way to accomplish this might be to simply listen to us and work with our problems."

The thought of having to work the research through university-based RAs (often a stated requirement of official grant application criteria) and community-based researchers was met with considerable disapproval and led to the stipulation that only ADEA community-based researchers could provide the knowledge and insights being sought through *par* and that outsiders would have no way of gaining the people's cooperation or trust because they are simply not of the people. The *par* process is currently progressing with the help of a team of six Kondh-Panos community-based RAs. Engagement of university-based RAs has been primarily at the level of literature searches and cataloguing of materials.

Methods of data collection during the course of the research process have been geared toward opportunities to speak together and individual engagements (e.g., attempts at one-on-one interviews with key informants) have been confined to elder narratives (*bakhanis* or stories/narratives with historical-political-moral import for movement consciousness). For example, on requesting a meeting with village leadership concerning land-struggle related learning engagements, the entire village was present (more than fifty individuals—children, men, and women, old and young alike) and while leaders and elders took the lead in sharing, all members participated freely in sharing their perspectives/analysis, stories, and factual information. The place for individual interviewing in *par* exercises is not likely to resonate with a people's PAR process in this context. More is shared on these collective research engagements in the following text in relation to knowledge mobilizations and movement politicization of data gathering exercises (*par*).

Data analysis (thematic analysis) has been met quizzically if not with some consternation (*par* attempts at collective analysis with the RA team or even village/leadership team members) or the role of member checks in ascertaining trustworthiness of data and interpretations (*par* processes/requirements) as participant communities when approached to help with such tasks, have often retorted, "we have spoken already about this," "do you not trust what we have already said?" and "we trust you, we know you." Or when it came to data analysis and coding/thematic developments, one community-based RA put it quite succinctly, "What are you doing to us?" The idea of blending/sorting/coding across people, villages, perspectives, and specific words shared over the course of data collection appeared to be viewed as an act of dismemberment. The integrity of what had been shared was lost in this act of data analysis, a protocol of *par*. Analytical re-presentations were in danger of mis-representing or worse still, missing a collective political message (a collage impact/meaning) through such analysis (blending down)—so rather than collapse, data analysis was to be more about expansion and connection of multiple whole expressions (pieces of data) that when taken together, would convey the desired meaning(s).

Choice of sites of *par* engagement were not necessarily the appropriate choices in relation to a movement politics and people's PAR, as the criteria defining such decisions were quite different. While *par* protocols would, for example, emphasize purposive sampling and snow-ball processes to lead to sites and prospective research participants in calm surroundings (to affect good interview prospects/rich data with due ethical consideration for participant safety and environmental appropriateness for same), people's PAR favored research/active sites that were embroiled in critical incidents (also an element of *par* possibilities with risk-ethical ramifications that could dissuade such site/circumstance selection) around land conflicts and tense engagements with law enforcement officers or with the lower reaches of the state-administrative bureaucracy.

When it comes to research budgeting, while *par* budgets involve necessary allocations toward in-university expenditures (e.g., for student GRAs, teaching releases, and conference presentations/travel for dissemination, often at other universities), people's PAR places an emphasis on the need for and maximal impact of resources in the research site and for the advancement of the people's movement process. For example, while a *par* budget would minimize money spent on, say, the purchase of grain for village grain banks (as acknowledgment for people's participation in the research process) by restricting these purchases to a delimited set of participant villages (half a dozen perhaps), a people's PAR position on this would be to extend this facility to include all 120 ADEA movement villages as the people speak, act, and belong as one. Plenty of opportunities were discussed that some times cohered in terms of meeting *par* requirements and a people's PAR as in the case of transportation, where for instance, bicycles and a motor bike were needed for research team mobility in hilly rural terrain while they could also be used to enhance mobility/communication between movement villages for leadership gatherings around pressing movement issues that could also be of *par* interest, not to mention that several people have availed themselves of these simple modes of transportation in medical emergencies that would otherwise have meant days of walking in some cases.

Research and Knowledge-Learning-Action Mobilizations: *par* and People's PAR Connectivity?

The research process around *par* provided plenty of opportunity for the ADEA participants to exercise the importance of a people's PAR in relation to knowledge-learning and action connectivities, both, during and part of *par*-specific data processes and after *par* episodes were brought to a close on any given visit or research moment. In fact, given a *par*-bounded commitment to a priori ethical and methodological protocols, the researcher or principal investigator (PI) was often surprised by demonstrations of movement learning and action incidents that confounded the boundaries between a *par*-focused episode and an emergent people's PAR engagement that simply picked up from where academic research had left off, so to speak. A few illustrative examples are shared here in relation to these eventualities.

On one occasion the PI and community-based RA's visited a Kondh village in the evening, something typically done after the men had returned from shifting cultivation in the hills, with the idea of a *par* focus on engaging people around songs on movement-related political themes pertaining to movement objectives concerning land, forests, and the dynamics of domination-resistance. Some sixty people (men, women, and children of all ages) assembled in the village square and sat around the *mandap* (raised platform). RA team members went around to each hut inviting people, while a young man arranged for an electric bulb ("rural electrification," he had commented with a grin on his face). After a formal introduction by the village leader (member of ADEA leadership) and a brief welcome and explanation around the purpose of our visit, we were treated to a festival of political/other songs by a group of women, the lead singer being an elderly woman who also played a *sarangi* (stringed instrument). As the evening progressed, the songs became more inspiring (RAs were taking notes on key songs and chatting with some people in the gathering about the significance of some of the songs) while I handled a recording device and sat with the village leaders and elders on the *mandap* as an honored "guest." Two hours later (approximately 10 p.m.), the following song was sung out and old and young alike wept as the singer pointed to the crowd repeating, "we weep together, young and old as you can see, *atma kanduchi*" (our souls weep) when we sing these songs and remember:

> In olden times oh brothers and oh sisters
> In the time of the British rule, the Britishers used our grandparents
>   like servants and beat them severely to make them work
> During that period oh brothers and sisters, the revenue collectors
>   came and took the measurement of our lands and paddy fields
> They said, "you will be given land, paddy fields and dry land"
> We went to work even when we did not have anything to eat
> But when work was done
> Our land and paddy fields and dry land were transferred to the
>   rich people and the big people, the outsiders
> From then we lost our way, from then we are hopeless...
> That is why we cleared up the mountains where the monkeys
>   lived and we started working there
> We cleared up the mountains where the tiger lived
>   and worked there...
> We struggled under the sun and rotted in the scorching heat
>   as we labored...
> We drank sour porridge and labored on, we became one with
>   the rock...
> Our rotting in the scorching sun was in vain as we did not get
>   enough crops from the *bagara* (land) on the hills...

Emotive in its appeal and powered by the strength of the suffering and injustices endured by ancestors and their contemporary plight, the village leadership took up the theme of usurpation and dispossession, as a Kondh leader stood up and began an impassioned speech:

> we are laying a claim on the government who is supposed to serve all the people of this land. We are demanding a place for ourselves—we are questioning the government and asking them to help us develop our land using our ways. ADEA's idea is that our livelihood should be protected and our traditional occupations and relationship to the land and forests be protected in the form of community control over our land and forests in our areas and this is our understanding of our constitutional rights too. There is no contradiction... If they can help the *shaharis* (moderns/urbanites) destroy the forests, then they can and should help us to protect it and listen to our story too.

The *par* session propelled several such speeches and ADEA related exhortations that went long in to the early hours of the morning, well after we had excused ourselves to make the journey down the hill in the dark, despite the generous invitations to stay the night. Where a *par*-related data gathering opportunity and process had "stopped," movement leaders mobilized people's knowledge from song and sense of history, taught about the ADEA and its directions and several days later, engaged the people of this village in a collective land-action involving demonstrations and sit-ins at the *tehsil* office (local administration)—a people's PAR took over where *par* had left off.

Similarly, *par*-related data gathering around people's history, culture, and ADEA movement aspirations formed the basis of the first issue of a people's research sharing journal, *Arkatha* (our talk/voice), as part of popular dissemination of research results (the official *par* process). Released before local government elections (*panchayat* institutions), *Arkatha* (1000 copies distributed on a per village basis to the 120 ADEA village region and surrounding *panchayats* [local administrative units/level of people's/village government]) triggered unprecedented popular excitement as people read materials written by them and about themselves and their ways. ADEA villages took it upon themselves to initiate reading circles around the issue and used it as a platform to promote critical debate pertaining to the elections, candidates, and party positions. Several dominant party candidates were defeated in the elections in these eight *panchayats* and a significant number of people's leaders were elected in the region. These elements of people's PAR extended movement dynamics initiated by *par*. Relatedly, the CRDS initiated through *par*, has now grown to include several initiatives addressing atrocities committed against Panos/Dalits and Adivasis in the region. The center provided logistic and research support to the National Commission on Dalit Human Rights (NCDHR) during ethnic-religious violence in the Kandhmal districts in 2006–2007 and the current round of related violence in October 2007. This process of documentation has been instrumental in human rights litigation and in demonstrating the reluctance of the state government and law enforcement agencies in taking decisive action to stop the carnage against Dalit Christians and Hindu sympathizers and neighbors who tried to prevent the violence.

Another example pertaining to *par* and people's PAR connectivity and possible political extensions is with respect to the emergence of subaltern networks (Kapoor, 2009). The *par* process set out to explore and understand movement-to-movement interaction by convening a gathering of fourteen subaltern movement/people's organizations in South Orissa, including different Adivasi groups, fisher-folk unions on the coast, rural women's organizations, landless peoples/peasant groups, development/mining displaced social movement organizations, and Dalit organizations. While *par* documented learning in movement gatherings for research purposes, the same gathering has initiated a process of movement networking that has resulted in joint statements of solidarity to address state-market led land and forest invasions in to Adivasi-Dalit-peasant communities in the region in the name of development and progress and is being followed up with a statement for possible joint actions in a translocal politics; that is, *par* has once again provided the basis for a people's PAR process that politicizes the research process initiated by *par* in the interests of subaltern movements and translocal movement activism.

In the final analysis, processes of data collection and joint analysis for *par* encourage movement introspection, sharing and gathering through the various data-related engagements constructed by the deliberate and partially planned attempt to accomplish research objectives and answer a priori questions/issues being raised by the research. No leadership gathering or village gathering for *par* purposes was left politically untouched by the *par* process, despite the academic commitment to abide by an apolitical, noninvasive, and risk-free participatory case study approach and to subscribe to a typical qualitative disinterest in the political dimension (over and above epistemic politics to considerations of political-economy and material struggles) of doing research in marginalized and politically charged contexts of subaltern movements. The political import of the research itself promoted reflection, reflexive conversations, and unintentionally galvanized movement villages through the interest in a line of collective inquiry that was somewhat germane to the harsh predicaments of people's lives. Unsurprisingly, the tidy plans of a *par* process committed to the organized excavation of people's lives for official knowledge (e.g., of some import for policy) or critical navel gazing exercises in academic menageries or university classrooms that are perhaps theoretically engaged but contextually disconnected, was often derailed by a people's PAR process embedded in a movement dynamic that demanded an urgent flexibility, as an unpredictable research context often required investigative attention as in the sudden eruption of caste-ethnic-religious violence just referred to. The video cameras were abruptly tuned to a set of unexpected movement-related events—such are the engagements with a people's PAR concerned with subaltern social movements. Once research that is participatory is effectively launched (even *par*) through extensive community participation among organized people (as in the ADEA movement scenario), "action" is difficult to contain (Feagin & Vera, 2001, p. 177) as a people's PAR could drive events beyond what is envisaged in a typical qualitative *par* process.

Lived-Theoretical Constructions and Theoretical Hearing Impairments

People's PAR relies on reflection (including various academic conceptions of the theoretical) that emerges from, returns to, and emerges from lived realities in a specific context of engagement. Versions of *par*, on the other hand, often rely on academic theoretical constructions in terms of a priori usage and or emergent usage in conjunction with data collection and analysis as linked processes or postpriori, in the search for grounded theorizations for instance, that are still contained, referenced, and/or influenced by a theoretical address in the academic repositories of accumulated socioeducational knowledge. Theoretical pre and postconceptions in *par* are confronted by the lived-theoretical conceptions of a people's PAR in movement. Consider the following conceptualizations and related descriptive-analytical-normative-reflective lived-theoretical propositions posited by various participants along the course of the *par* in to "Learning in Adivisasi social movements" in Orissa, for instance, in relation to processes (and related concepts) of accumulation by dispossession, social oppression, agents of domination, structural constraints/domination (hegemonic impositions), and dominant-group perceptions of Adivasis/Panos peoples (all indicative of a vibrant subaltern political consciousness):

> Life in the town is not created for our type of life. The people of *shahar* (city) will never think about us. They would rather enjoy life from the benefits that come from the forest and mountains, like water and forest products. They tell us they want to modernize, make machines, and industries for themselves. To do this, they are doing forcible encroachment of our land—they are over our hills and stones. They are coming quietly to our forests and hills and in secrecy they are making plans to dig them up and destroy them (mining). Not only this, they are diverting our water to the towns (dams) for their use. They are making dams and water reservoirs, where our villages are to be submerged and we have to leave the land and become landless and homeless. We have become silent spectators (*niravre dekhuchu*) to a repeated snatching away of our resources.
>
> The *sarkar* (government) is doing a great injustice (*anyayo durniti*) and is involved in corrupt practices... and the way they have framed

laws around land-holding and distribution, we the poor are being squashed and stampeded in to each others' space and are getting suffocated (*dalachatta hoi santholito ho chonti*). This creation of inequality (*taro tomyo*) is so widespread and so true—we see it in our lives...

They have the power of *dhana* (wealth) and *astro-sahastro* (armaments). They have the power of *kruthrima ain* (artificial laws and rules)—they created these laws just to maintain their own interests. (Kondh Adivasi elder, field notes)

The government and the companies come and take away truckloads of bamboo.... When they take truckloads of sal timber, bamboos and the paper mills exploit this place for their business—how can they say the Adivasis are destroying the forests when they are the ones doing this? When the Adivasi depends on the forest for their life, the *vyavasahi* (business people) and the government are destroying them for profit (*labho*). (Kondh Adivasi man, field notes)

Whenever we have tried to assert our land rights, we have been warned by the upper castes, their politician friends, and the wealthy and have faced innumerable threats and retaliations. The *ucha-barga* (dominant castes and classes) will work to divide and have us fight each other till we are reduced to dust (*talitalanth*). (Kondh Adivasi adult, field notes)

The *sarkar* (government) and their workers think that we Adivasis do not know anything and we are good for nothing, that we are weak and powerless and will not question them if they treat us unjustly... to the government we are of no significance (*sarkar amar prathi heyogyano karuchi*). They are selling our forests, they are selling our water and they are selling our land and may be they will sell us also. (Kondh woman leader of the ADEA, field notes)

Or similarly, consider some of the following comments shared in relation to academic notions of agency and resistance/resilience, movement teleology and strategy, hope, strategies/directions for change, and the role of learning/education in social movement action:

Who wants to go to the city to join the Oriyas and do business and open shops and be *shahari* (city/modern) if they give you a chance or to do labor like donkeys to get one meal? Even if they teach us, we do not want to go to the cities—these are not the ways of the Adivasi. We cannot leave our forests (*ame jangale chari paribo nahi*). The forest is our second home (after our huts). We have a deep relationship with our forests (*jangale sahitho gobhiro sampark*). There is no distance between our homes and the forest. (Kondh Adivasi village elder, field notes)

We are the real protectors of the forest (*ame jangalo surokhya kariba lokho*) because we are the ones that have always depended on the forests for our wellbeing. For us all plants and animals are equally valuable... our forests are our history and our culture (*amor jangalo, amaro itihas ote avom a thi amoro sanskriti*). (Kondh Adivasi woman, field notes)

Ekta Abhijan (ADEA) stands on a root called unity (*ekta*) and the promotion of unity will always be the primary requirement...The artificially created sense of difference, divisions, and *jati-goshti* (caste-class feelings) need to be destroyed. Our *dhwoja* (flag) is unity (*ekta*) and we have to fly it high (*oraiba*). The flag that the ADEA flies is of the people who have lost their land and their forests and who are losing their very roots. (Panos ADEA leader, field notes)

The ADEA is there to fight collectively (*sangram*) to save (*raksha*) the forests and to protect our way of life. The ADEA is a means of collective struggle for the forest (*ame samastha mishi sangram o kariba*). We are all members of the ADEA and our struggle is around *khadyo, jamin, jalo, jangalo o ektha* (food, land, water, forest, and unity). (Kondh Adivasi woman, field notes)

We are giving importance to land occupation *(padar bari akthiar)* and land use *(chatriya chatri)*. We are now beginning to see the fruits of occupations. Before the government use *se anawadi* land (unused state land) to plant cashew, eucalyptus or virtually gives the land to bauxite mining companies, we must encroach and occupy and the put the land to use through our plantation activities and agricultural use. This has become our knowledge through joint land action. This knowledge is not only with me now but with all our people—what are the ways open to us—this is like the opening of knowledge that was hidden from us for ages. (Kondh Adivasi male, field notes)

Even though some people still say that this is our destiny (*bhagya*), most people today because of the ADEA action, would challenge this idea of *bhagya*...we have to teach each other (*bujha-sujha*)...we organize workshops and gatherings and have created a learning environment—we have been creating a political education around land, forest, and water issues and debating courses of action. (Kondh Adivasi man, field notes)

If the government continues to control land, forests and water, we will have depended on for since our ancestors came, then through the ADEA we will be compelled to engage in collective struggle (*ame samohiko bhabe, sangram kariba pahi badhyo hebu*)...I think this movement should spread to the district level. The organization is always giving us new ideas (*nothon chinta*), new education (*nothon shikhya*), awareness (*chetna*), and plans (*jojana*). We believe this will continue (*ao yu eha kari chalibo amaro viswas*). (Saora Adivasi ADEA leader, field notes)

The lived-theoretical constructions of a people's PAR process can fall on "theoretically hearing-impaired ears" (Baxi, 2000, p. 45) of a *par* process, especially *par* variations that are fraught with epistemic slippages toward pseudoscientific propositions in an attempt to conform to the ruling relations of funded research to, perhaps, secure respectability, rewards, and recognition. The compulsion to conform to or move from and to academic theoretical knowledge places *par* and dominant approaches to socioeducational research in general, in a position where: (1) there is an inability to give voice to human suffering, manifest or latent, in oppressive structures of the modern state, civil society, and the market, even when it deals with movements of the repressed (e.g., subalterns) as "cognizing human suffering is to take a position towards it and this disturbs the canon of value-neutrality in social theory construction" (Baxi, 2000, p. 37) (as is evident in *par* emulations of the same scientific striving); (2) human suffering may enter theorizing but as an "assemblage of social facts which need categories of description and explanation but the ideal of positivistic social theory is to look at the facts of suffering as no more than raw materials for epistemic construction" (Baxi, 2000, p. 37)—an analytic standpoint that "plays a part in the effort to silence suffering coming to voice" (p. 39); or (3) the production of a narrative of suffering that commoditizes it by way of professional appropriation of it (Kleinman & Kleinman, 1997) and/or the production of suffering as a spectacle (Debord, 1990).

The danger of theoretical hearing impairment produced by blind adherence to the scientific canons of social theorizing and their associated productive political interest vis-à-vis ruling relations of capital, is made evident in Upendra Baxi's (2000) critical rendition pertaining to preferred social scientific explanations/representations provided for the Bhopal victims-activist movement against Union Carbide with respect to the disaster of 1984, where in social movement theorists and academics described this as a mere movement for compensation, deaf to the "roar from the 47 tonnes of MIC impaired lungs of 200,000 human beings" shouting the popular slogan "Union Carbide *ko phasi do*" (hang Union Carbide) or essentially a movement demand for the decapitation of a Fortune 500 company (p. 45). It follows that theory and theorizing with respect to subaltern social movements like the ADEA is better served by theoretical reflexivity on the part of *par* and a closer adherence to the integrity of lived-theoretical constructions developed by movement actors themselves. A priori theoretical commitments ensconced in research proposals for funding are subsequently suspect, to say the least.

Similarly, when it comes to questions of theory and politicization in activist versions of PAR (scholar-activisms and PAR as opposed to conventional *par*) committed to historical materialism, activist theoretical impositions in the name of a political project can also lead to another form of critical theoretical hearing impairment candidly acknowledged by Orland Fals-Borda in reference to his work in Columbia in the 1970s (Fals-Borda, 1979, p. 49) where groups fell to the historical dogmatism of "mimesis"; that is, historical materialism as a theoretical heritage of activist researchers and intellectuals committed to the development of a science of the proletariat (or legitimization of popular knowledge in to scientific knowledge) found little resonance with the masses when it came to taking the initiative to continue with the intellectual inquiry at which point, "impatient action researchers and their intellectual allies were forced to inject their own definitions of popular science in to the context of reality" (Rahman, 1985, p. 12). Knowledge was thereby placed at the service of popular interests, "but such knowledge did not derive from the objective conditions of the proletariat as would have been theoretically more correct" (Rahman, 1985, p. 12). In response, Rahman suggests that, "People can not be liberated by a consciousness and knowledge other than their own, and a strategy such as the above (where a vanguard body with a supposedly advanced consciousness liberates the people) inevitably contains seeds of new forms of domination" (pp. 14–15). When it comes to theoretical knowledge and political engagement, Rahman makes it clear that it is "absolutely essential that the people develop their own endogenous process of consciousness raising and knowledge generation and this process acquires the social power to assert itself vis-à-vis all elite consciousness and knowledge" (p. 15). These are but additional observations around the significance of listening to people's knowledge and theorizations in subaltern movements like the ADEA; listening with ears that are not theoretically compromised by virtue of a dogmatic critical politics of liberation.

Research, Politicization, and Subaltern Social Movement Spaces

In an era of higher education where research bids for funding predominate, an "investor rationality prevents focus on repressive structures that cause human suffering—research has become increasingly hi-tech, capital-intensive and tied to examining the structures of capitalist hegemony" (Baxi, 2000, p. 39). Given these controlling influences on the structure of knowledge production in social and educational research, it is unlikely that a people's PAR challenging these sociopolitical vectors is likely to gain academic respectably and neither should this, as it ought to be clear by now, be an aspiration as the political relevance of a people's PAR is derived primarily from its critical location outside ruling relations. However, if academics with a critical social agenda are to find even a minor role in the democratic struggles of groups like the ADEA, there is some political hope and grounded possibility in versions of *par* that seek to get closer (participatory) to marginalized communities and their politics; that is, despite striving to be consistent with the canons of academic research within capitalist, Eurocultural relations of rule (in fact, while being consistent with this politics), *par,* if attempted in radicalized and organized settings such as within subaltern social movement milieus, will be subjected to a politics of the margins (at which point, academic researchers could chose to become willing hostages in the interests of democratizing research relationships) as movement actors build on *par* conversations and resource supports in the interests of continuing to make their history.

In the final analysis, there is and will continue to be plenty of scope and need for *par* and people's PAR, as neoliberal globalization, modern development, and displacement/dispossession adds to the ranks of the marginalized (Davis, 2006) and more significantly, continues to be met by a proliferation of subaltern movement articulations (Martinez-Alier, 2005), "whose presence in the cultural and political fields opens new possibilities for the ways we live our lives and define ourselves" (Carroll, 2006, p. 234).

Notes

1. This term comes from Antonio Gramsci's (1971) use of "subaltern" (p. 55) and "subaltern consciousness" (pp. 325–326) in relation to the Italian peasantry and is being used here to refer to Adivasis, low/caste Panos agricultural labor, sharecroppers, smallholder peasants, artisans, shepherds, and migrant landless labor working in mines and plantations. The term also alludes to the dialectical relations of superordination and subordination that define social relations in hierarchical social formations, keeping in mind Guha's (1982, pp. 5–8) observation that there are ambiguities inherent in the concept when applied to the Indian context.
2. According to Balakrishnan Rajagopal (2004), some 33 million people have been evicted from their homes due to various development projects (e.g., dams) in the postindependence era in India, "development refugees" who are the product of "development repression" that is not considered a human rights violation—"It is unfortunately true that violence (economic violence) committed in the name of development remains 'invisible' to the human rights discourse" (p. 195).
3. The author acknowledges the assistance of the Social Sciences and Humanities Research Council (SSHRC) of Canada for this research into "Learning in Adivasi social movements" in India through a Standard Research Grant (2006–2010).
4. By people's PAR, I mean a mobilization of subaltern knowledge (Fals Borda, 1979; Rahman, 1985), and subaltern pedagogy (Freire, 1970) working together in the interests of a subaltern action/politics of decolonization (Grande, 2004; Smith, 1999) or action research that is participatory and participatory research that unites with action for transforming reality (Fals Borda, 1979). Alternatively, as per Hall's (1979) identification of key components: a method of research involving full participation of the community, as a dialogical educational process and as a means of taking action for change.
5. While Adivasis constitute 8 percent of the Indian population (approximately 80 million people belonging to some 612 scheduled tribes), they account for 40 percent of development-displaced persons and in Orissa (home to 62 tribal groups numbering 8 million or more people) while making up 22 percent of the population, they account for 42 percent of development-displaced persons (Fernandes, 2006, p. 113).

References

Baxi, U. (2000). Human rights: Suffering between the movements and markets. In R. Cohen, & S. Rai (Eds.), *Global social movements* (pp. 33–45). New Brunswick, NJ: Athlone Press.

Carroll, W. (2006). Marx's method and the contributions of institutional ethnography. In C. Frampton et al. (Eds.), *Sociology for changing the world: Social movements/social research* (pp. 232–245). Halifax, Canada.

Davis, M. (2006). *Planet of slums*. London: Verso.

Debord, G. (1990). *Comments on the society of spectacle*. London: Verso.
Fals-Borda, O. (1979). Investigating reality in order to transform it. *Dialectical Anthropology, 4*(1), 33–56.
Feagin, J., & Vera, H. (2001). *Liberation sociology*. Boulder, CO: Westview.
Fernandes, W. (2006). Development related displacement and tribal women. In G. Rath (Ed.), *Tribal development in India: The contemporary debate*. New Delhi, India: Sage.
Freire, P. (1970). *Pedagogy of the oppressed*. New York: Continuum.
Gramsci, A. (1971). *Selections from the prison notebooks*. London: Lawrence & Wishart.
Grande, S. (2004). *Red pedagogy: Native American social and political thought*. Lanham, MD: Rowman & Littlefield.
Guha, R. (1982). *Subaltern studies: Writings on South Asian history and society*. New Delhi, India: Oxford.
Hall, B. (1979). Knowledge as a commodity and participatory research. *Prospects: Quarterly Review of Education, 9*(4), 393–408.
Kapoor, D. (2004). Popular education and social movements in India: State responses to constructive resistance for social justice in India. *Convergence, 37*(2), 55–63.
———. (2007). Subaltern social movement learning and the decolonization of space in India. *International Education, 37*(1), 10–41.
———. (2008). Globalization, dispossession and subaltern social movement (SSM) learning in the South. In A. Abdi, & D. Kapoor (Eds.), *Global perspectives on adult education* (pp. 100–132). New York: Palgrave Macmillan.
———. (2009). Adivasi (original dwellers) "in the way of" state-corporate development: Development dispossession and learning in social action for land and forests in India. *McGill Journal of Education, 44*(1), 55–78.
Kleinman A., & Kleinman, J. (1997). The appeal of experience; the dismay of images: Cultural appropriations of suffering in our times. In A. Kleinman, V. Das, & M. Lock (Eds.), *Social Suffering*. Berkeley: University of California Press.
Martinez-Alier, J. (2005). *The environmentalism of the poor: A study of ecological conflicts and valuation*. New Delhi: Oxford.
Rahman, M. (1985). The theory and practice of participatory action research. In O. Fals Borda (Ed.), *The challenge of social change*. London: Sage.
Rajagopal, B. (2004). *International law from below: Development, social movements and third world resistance*. Cambridge: Cambridge University Press.
Smith, L. (1999). *Decolonizing methodologies: Research and indigenous peoples*. London: Zed.

CHAPTER FOUR

When Research Becomes a Revolution: Participatory Action Research with Indigenous Peoples

CORA WEBER-PILLWAX

Preface

My reflections on Participatory Action Research (PAR) have been long and, in many ways, a test of personal endurance in all aspects of my being. In positioning myself as an Indigenous researcher in a world of contemporary academia, however, I need to go back first to my childhood where I lived and learned through an Indigenous research methodology (IRM) or paradigm. It was many years later, after having entered the professional world as a formal educator/teacher, that I ran face first into the concrete realization that my IRM was not effective as a research paradigm when I was engaged in the world of non-Aboriginal mainstream Alberta, Canada. The reasons for this have continued to unfold through the years. In looking back, I see that it was at that moment and with that realization that I was challenged with finding another way to live and work within the institutions of mainstream Canada. It was clear to me that, without a different approach to dealing with the impacts of mainstream social institutions on the social and personal lives of Aboriginal persons and communities, we would not survive as Aboriginal peoples.

I had left my family and community at a young age for high school and university. During those years, I incorporated the knowledge of mainstream institutions into my own repertoire of Indigenous knowledge. There had never been an opportunity for me to contribute my Indigenous knowledge to the formal institutions of education but I had not seen that as an issue. This remained my perspective until one day I noticed that I was teaching unilingual Cree children how to read in the English language without any resources or curriculum to assist me. In approaching the ministry, the university, and the school board officials, I had expected that they would respond eagerly with support and encouragement in the form of resources and expertise, as per their educational program manuals. I returned home with nothing. It was then that I began to see the world in terms of politics and power, and I let myself be driven by the energy of my own anger to engage in this environment. Many years later, as a graduate student, I realized that the people to whom I had turned for assistance had not possessed even the shadow of an idea that Indigenous knowledge systems existed and were, in fact, being lived out in the embodiment and agency of individual and collective Indigenous persons and communities. Although today the existences of Indigenous knowledge systems has been and continues to be acknowledged formally, especially in international forums, there remains a basic and profound societal and professional ignorance in relation to how such knowledges are to be incorporated into or given consideration in the administration and programming of contemporary schooling systems. Aboriginal education—except in cases of First Nations control where similar issues are compounded by political and financial ones—operates today primarily as an educational approach directed by provincial policies of curricular requirements for infusion/inclusion of bits and pieces of knowledge from various Aboriginal peoples in a process that decontextualizes and isolates Aboriginal people from their human identities, community spaces, and histories.

In 1980, I was called to participate in an educational effort that was grounded in Northern Alberta communities and offered opportunities for Indigenous research methods and political action to change the way schooling was happening in the isolated Aboriginal communities. At the time I did not know it, but that was my first engagement with the processes I came to identify later with PAR.

Beginning Research

I began my journey as a child who loved the world passionately, and I look forward to ending my journey in the same way. I remember myself as a child when every day the world and I just were. Then I remember myself being overwhelmed at the immensity of beauty in the world around me. And finally, I remember myself in later years as a youth consumed by an unbearable aching because I was not able to take in, understand, or fully experience that beauty. As I grew, I had experiences that led me to a consciousness of myself as an entity entirely separate from and simultaneously, inseparably connected to the world outside of myself. I slowly learned how to think, speak, and live in relationship with that world. Learning how to live in relationship with the world was not only about for respect and ethics or morals, but it was also about survival of me as both person and people.

In this process of growing up with theoretical and practical learning, I was slowly becoming a researcher in the Indigenous *sakaw,* or bush system of education and scholarship. This was participation in life through sustained observation, examination, inquiry, data collection, and content and process analysis. I systematically carried out experiments with samples and test cases over time, maintaining systems of record-keeping and documentation that I knew were expected of me as an independent and responsible learner. I was immersed in a research practice that was tens of thousands years old, and I was being mentored by qualified researchers who had earned their credentials before me through their own years of effective research engagement, as evidenced in observable and measurable beneficial outcomes in peoples' lives. In my understanding, this was and continues to be an IRM that guides the ways that Cree and Métis persons engage in the creation of new knowledge in support of their own transformation and toward the positive development of individual and collective identity and autonomy. These were necessary and vital processes for Indigenous being. After I had grown up, and discovered that I needed some additional skills, a different approach, a different mindset to support my research and work where I found myself at the interface of Indigenous and non-Indigenous social realities, I used the principles of my own IRM to find a methodology that was connected with the knowledge systems and historical experiences of mainstream Canada and yet could be grounded in the historical and contemporary realities of northern Cree communities in Alberta. This was PAR.

The intent of this piece of writing is not to put energy toward translating or transposing this meaning of PAR into the contemporary world of academic research and commentary about PAR, nor to what PAR looks like in that world. However, I hope this work will contribute to the forward impetus of PAR as an investigative methodology that is "a rite of communion between thinking and acting human beings, the researcher and the researched," where academic institutions give space to "some sort of down-to-earth collectivization in the search for knowledge" (Fals-Borda, 1997, p. 108). The approach indicated here offers hope and positive impacts from research within Indigenous communities and with Indigenous peoples. I acknowledge with gratitude this opportunity to speak about PAR and how it is situated within the research world where I live and work.

Introduction

PAR has gone through a number of significant name and practice changes over the past decades since Fals-Borda gave it an international profile in the late 1950s and early 1960s through his work in Colombia. In his obituary, Fals-Borda is credited in his academic role as sociologist at the National University of Colombia (Bogota), with developing the "methodology known as participatory action research, combining research and theory with political participation" (Gott, 2008). At that time and in that place, peasant and Aboriginal populations of his country were losing an ongoing and constant struggle for land rights with wealthy landowners supported by government and its bureaucracies. Fals-Borda believed that it was the responsibility of the researcher not only "to examine the social reality of the country, but to try to remedy the injustices that the research uncovered" (Gott, 2008). He had left his position at the University of Colombia because he found himself facing several intellectual dilemmas, one of which was the way that he saw his institution ignoring the theory and practice dialectics, supporting "learning by rote without relating to surrounding social and cultural realities," (Fals-Borda, 1997, p. 108) and believing science to be a knowledge distinct and separate from popular knowledge or people's knowledge. He left because he could not accept the apparent opposition between thought and action, or the imposed dichotomy of subject-object in research relationships. He left the university to engage in life-experiences (*vivencia*) as "authentic participatory practices in research and action" (p. 108).

In this work, I am referring to PAR as a research methodology that is a means of support for the intellectual and spiritual revolution of Indigenous peoples in Canada. The focus of PAR as an ongoing iterative process that involves and engages spirit, intellect, emotions, and physicality of human beings is one that is fully complementary with the research processes that were developed and continue to be followed by ancient and Indigenous peoples. Many contemporary Indigenous researchers, including Brown and Strega (2005), Ermine (1995), Hampton, (1995), Meyer (2002), and Steinhauer (2008), have described and otherwise articulated these research processes as particularly derived and intrinsically connected to the original sources of their own Indigeneity.

As a contemporary Indigenous researcher, I recognize PAR as a methodology that raises few inherent incompatibilities with the frameworks represented by my own Indigenous research methodology. However, I have to add also that PAR, in its purest form, is rarely adopted by researchers, including Indigenous researchers, to frame their work and this speaks to a broad recognition of the difficulties and the complexities that are embedded within the use of PAR. The successful engagement of communities and persons within the processes of PAR as a research methodology leads, without question, to an unleashing of power or energy of which the quality and degree is unpredictable, and which cannot be manipulated or controlled by the researcher if it is to be recognized as PAR.

PAR has, over the past couple of decades, been reconfigured, reshaped, and renamed as community-based research, action-research, participatory learning, participatory evaluation, appreciative inquiry, and praxis intervention. In the context of research with Indigenous peoples, these terms can represent safe ways to divert (or divest) and/or swallow up the potential power of research to serve the people, who are necessarily both subject and object of ongoing research. In my reading and interpretation of my preferred works of Fals-Borda (1997) and Paulo Freire (1970), PAR was never intended to be used as a research methodology outside of research conducted with a community of people who desire and are willing to work for some transformation or change in self and/or personal/collective environment. The intent of such research is the attainment of individual and collective researchers' empowerment, as evidenced in the visible and measurable aspects of such transformation. This is contrary to a positivist position that sets the researcher and the researched on opposite sides of a subject/object dichotomy, with the presupposition that researchers will be distant and dissociated from the researched, being both consciously subjective and consciously objective in relation to the researched, at every point of the research process through time and space. However, PAR is not functional outside of the consciousness of each and every individual member of those who are engaged in each and every one of the research processes.

Participants engage in the evolving processes of the research knowing that one, some, or all of them will benefit or lose to a lesser or greater degree in the outcomes of those research processes, and they also know that these outcomes are based on their actions and decisions as individuals and/or as group. Each participant enters into the experience of PAR knowing that it holds a promise of transformation, and knowing also that once having begun such a personal engagement, there is no turning back from that process of revolution or turning back around of self/environment. To engage in such transformation can and does call for a giving up of the personal. For many PAR researchers, this choice is not represented by the dichotomy of self and collective; it is merely a revolutionary experience of self as collective. For others, this dichotomy is evident, either from the beginning of the research project or as a realization somewhere along the process. In either case, PAR researchers do have to choose whether they will give themselves up to be a part of the whole or whether they will engage in some degree of participant observers. The dichotomy will sometimes be experienced by a researcher as a breaking point in the research process. The researcher will feel compelled to pull back and to remain within the security of the dichotomy, separated from research participants (situated within a research context) and who then can be observed from outside of that research context, at will, by the researcher. This particular aspect of researcher choice in working within PAR is one that often goes unmentioned in projects that claim to be doing PAR; yet, the integrity of PAR rests on the conscious choice of the researchers to give up the personal for the collective and it is this point that distinguishes PAR and is a measure of its power for transformation and knowledge creation.

In the following work, I will refer to Indigenous lived experiences in the context of traditional Indigenous communities. This context is integral to my thinking of PAR as a means of revolutionizing the way that we as researchers carry out, understand, articulate, and evaluate our work with Indigenous communities as individuals and as collectives.

In an earlier work (Weber-Pillwax, 2003), I described how participation in research demands a certain level and type of personal commitment to undergo the process and I spoke about how, within the processes of research, we find ourselves in patterns from which we "create new meanings of ourselves and we experience ourselves as transformed by the whole process of participation and immersion within the patterns of meanings and experiences" (p. 37). This section of the work also drew comparisons between my research methodology and that of PAR as described by Fals-Borda and Rahman (1991), wherein they talked about (1) the "ontological possibility" of real popular knowledge (science); (2) "the existential possibility of transforming the researcher/researched relationship"; and (3) the need for "autonomy and identity" in the exercise of people's "power" (p. 37). In all of my work, the focus on the lived experiences of the participants presupposes the legitimacy and validity of personal knowledge and the existence and reliability of Indigenous knowledge. As I said earlier,

> It is not because I am buying into the "ontological possibility" of "popular knowledge" that I am choosing to recognize those elements of PAR. It is because I respect the overall approach which vests the power of knowledge creation within the people themselves and at the same time recognizes and addresses explicitly the political nature of the challenges inherent to research conducted in this manner. (Weber-Pillwax, 2003, p. 38)

Relationship is the critical factor in the development and establishment of respectful Indigenous research projects. There is always a relationship between researcher and area of research, but when the research topic is embedded within human beings, then methodologies or strategies must be designed within the cultural parameters and definitions of respectful relationships. Otherwise, the researcher treads the dangerous ground of attempting to mine resources that are held within an individual person. In focusing on the treasure as the goal of the research, a researcher may forget the significance of relationship with that person who bears a sacred trust for that treasure. Standard academic research paradigms tend not to recognize or incorporate into their models the complexity or depth of knowledge that is required for researchers to develop successful research projects with Indigenous peoples. This knowledge often flows from the principle that Indigenous forms of research involve the establishment of particular and relevant forms of relationship among all living things connected within the research itself.

Transformation is another significant principle active in my research and one that I read as highlighted by Fals-Borda (1991, 1997), although it is not often mentioned in contemporary literature about PAR. In PAR, research participants inform and teach each other about the different worlds of knowledge that each inhabits and/or holds within capacity and competence. It is a relationship that is intended to create and enhance opportunities for personal and mutual empowerment through the recognition, acquisition, and/or creation of personal and shared knowledge systems—knowledge that is multidirectional and multilayered, taking unpredictable forms. These particular forms are not necessarily recognizable and evident in exterior aspects and behaviors of persons. They are more often displayed as elements within the contents and processes inherent to relationships themselves, abstractions within abstractions, and therefore can be difficult to ascertain.

Within PAR processes, personal transformations occur in a variety of degrees and types. In an Indigenous context, these transformations and cases of personal empowerment often take place as outcomes of the gradual discernment and unwrapping of existing Indigenous knowledge structures governing relationships with the ancestors and the spirit world. The participation of the ancestors in the development of Indigenous personal and community knowledge is accepted and known to encompass more than resonant narratives and poetry that we relate to each other and to others as memorization and polemics for didactic purposes. Whenever the words we choose to use are accepted and validated by the experiences of the people, whether of past or present, and whether in ceremony or ordinary space and time, they become expressions of the autonomous and distinct nature of individual and collective identities.

PAR stresses the importance of autonomy and identity as indicators of people's power. In PAR, the autonomy and identity of the people are the measures of the power of the people—a high degree of autonomy and a strong sense of identity indicates a lot of power in the people, presumably over their own lives. From an Indigenous perspective, lived experiences are the measure and expressions of autonomy and identity. They are interconnected and interrelated in such a way that each affects and shapes the other. Lived experience shapes identity and gives reality to autonomy. Identity shapes lived experience and embodies autonomy. Autonomy gives meaning to identity and can determine the shape of lived experience. Indigenous peoples tend to see all lived experience as sacred since the human is a sacred being, and it is impossible to isolate identity from lived experience.

Regarding personal empowerment and PAR, it is not likely that any research would ever have been carried out successfully among the Northern Sakaw Cree communities without strong individual and collective senses of identity and autonomy among the participants. An individual cannot experience and interpret the meanings of identity and lived experiences in a state of partial confinement or restriction to culturally other definitions of time, space, and history. In other words, the individual's sense of empowerment must flow from knowledge systems that have evolved to enable him/her to create new meanings and new experiences relevant to Indigenous beings as Indigenous persons. Our lived experiences must be given expression in an explicit way if we are to understand and maintain autonomy over them. These expressions become our new histories as individuals and as peoples, becoming the content of our narratives of hope for our children and their futures (Weber-Pillwax, 2003).

Personal Reflections on Transformation through PAR

PAR is transformation of self and transformation of environment. As Indigenous peoples, we know when we enter into the process of engagement with our world. When it is authentic engagement, and we see our world being transformed, we know that we are also being transformed. With or without analysis, we know that we are also being transformed as we transform the environment.

The work of ceremonies is to try to support that process of transformation of self, so that every aspect of your being stays in balance as you are being transformed and as you transform your environment. The agency does lie in the person, in the individual; when you act upon your environment, you, in that sense, are transforming yourself. The environment is not limited by definition to the physical world environment; it also refers to the spiritual world, and to other planes of existence that are entered into through ceremony. Ceremony, then, maintains personal balance. It also enables personal engagement with other levels of existence, with other parts of the environment that an individual may not be able to physically see. Therefore, when viewed from this perspective, transformation in order to be total has to involve and engage all aspects of a person's being, of his/her individual humanity, of the individual's sense of personhood. Transformation impacts every aspect of the human being: the intellectual, the spiritual, the physical, and the emotional.

In thinking about the values and contributions that PAR can offer to Indigenous and other research, I am remembering the revolutionizing force that happens inside of the research participant as an essential part of the whole process. PAR inspires because it presents a learning experience; it is grounded in a theory that says when you are learning, you are alive.

Transformation of self and transformation of environment is the dyad of reality; the dyad of being for me, and for my environment; the dyad of being for every other person in my collective, my cultural grouping.

Because my research is me, as in it is not research outside of myself, it is very difficult to write about. It is difficult in the sense that while I am engaged (in the research), I cannot be writing. I cannot be engaged fully in the act of living, and simultaneously engaged fully in the act of writing that is actually creating some kind of image of my own life on paper. To write about myself engaged in the act of living is to create an icon of some sort so that a reader can "read" me and understand my life.

If I write for you as the reader, you will more likely be able to understand my people because I, as a trained academic, am more likely to be able to write in a way that is more likely to help you understand me and my people at home.

Indigenous children in schools are at the mercy of adult training processes, mainstream and non-Indigenous. The people who are there to teach and form these children have been trained by those same processes. They have not been trained to see the Indigenous beings of those children. Should I be writing to help in the training of those people so that they in turn can better train my children? The key question that has to be answered is: whose definition of "better" is to be the standard of measure?

The only possible way that I can ever hold onto my children and help them to become who they are—the being of themselves as humans—and the only way I as the adult responsible for their care and well-being can hold onto that for them, is to ensure that I can balance at home what is happening to them in school or I can start to control what is happening to them in school.

I am realizing that that is nearly an impossible task. As Indigenous parents, we cannot do either because the tentacles of oppression and destruction reach into every aspect of that child's being—my child's being. Therefore, I come back to my earlier statement: if I am not ensuring and if I am not taking care of my child's experiences at school, then the persons who control that schooling system have my child. Those persons have the spirit of my child in their hands and they have no idea what they are holding. They have no idea how to care for that child or that spirit. The sad thing is that they do not even know that they do not know how to do that. If I write to help us to understand each other, I am seen as this little icon of me that is represented by my thoughts on paper. While we engage in discourse and theoretical dialogue on paper, and I live my life in that meaningless connection, my children are dying.

As Indigenous researchers, we do not say PAR very often; we say Indigenous Research Methodologies (IRM), and we ground it in Indigenous communities and we talk about transformations as a natural part of research processes. Basically, I weave the principles of PAR in and through the principles of IRM and then in the colors, the threads that are going to be the warp and the woof of the weaving, you see Indigenous peoples represented—because it is not about research, it is about living. And IRM are about sustaining, supporting, enriching, enhancing, and uplifting life/living in a way that is transformative and based on transformation. PAR, in its very basic principles, can be applied to any people or cultural group anywhere because it also aims for transformation. The objective is transformation, change, creating new knowledge, uplifting, inspiring—all of those same terms apply and they are all based on transformation. That is PAR and revolution, in any language.

PAR as Everyday Revolution

Every day of our lives we are engaged with PAR as Aboriginal people because being engaged with life within the mainstream society is not us living within our own environment. Each morning, we step out of our own comfortable space and move into a space where we are constantly in researcher mode. We are constantly required to weigh, observe, assess, make decisions, and act. Constantly, day in, day out. I realize the argument is compelling to rejoin and say "oh well, we all do that." It is true that we all look, assess, make decisions, act. However, I think we do not all do that in a foreign environment, not in the way that that demand is placed on us as Aboriginal people. We come out of our own places and we go into a strange environment, very often unfriendly, very often hostile, and that is where we have to operate every day. We did not create the environment so therefore it is not a research environment that we create, and it is certainly not an environment that we manage or that we control. Yet we control ourselves every day in choosing to become involved in that environment. Every day when we get up and go forth into our day, we are choosing to involve ourselves in PAR. This statement stands in relation to people and the power of the word, critical thinking and critical actions (Freire, 1970)—all of those things which are a part of what guides us as we enter into our situation of PAR every day as Indigenous/Aboriginal researchers. We know PAR as a very powerful tool because we knowingly engage and we knowingly participate, we intentionally aim for transformation. We know these are the things that actually lead to positive changes and the deeper this transformation and the more it becomes part of our lives, the more of a revolution it is: a revolution in the sense that it turns us back into ourselves, into new and ancient ways of being.

That is being constantly in a PAR mode, just to survive. We started early to learn, some of us as preschoolers, some as teenagers, some as adults. Many of our people never did demand such effort of themselves; my mother was one such person. As a young adult, I realized slowly that there were many rules to learn about engaging with mainstream Canada: the language to use to communicate (in words and body movements), what was funny and what was not, what defined wit, puns, morals, jokes, measurements, clichés, fashions, literature, music, taste and foods, size and appearance, men and women, and the hundreds of tiny details that make up a culture and a people in every moment, every circumstance, and every space. And, in the midst of all these, I knew who I was and where I belonged.

When we as Indigenous persons enter into PAR, we want to learn something, we want to move ahead, we want to win something, we want to gain freedom, we want to gain the rights to be creative in our own ways, to create knowledge that is applicable to our own lives, we want to transform the societies within which we are engaged. We want to transform the broader society in the sense that we want our own society within that context to be a safe and comfortable place for us. It is not acceptance that we are looking for, it is our own space. PAR begins with the decision to do something—the decision that is founded on an intention to engage in transformation. The intentionality is part of the methodology. Without the intention, the methodology is meaningless because it will not result in transformation. Intentions are a critical aspect of research and are in place, stated or unstated, well before the physical activation of any work. If the intention is unclear, the impact of the research cannot be anticipated and is more likely to go unnoticed. Even in the best of circumstances, a researcher can only know within limits the impact of his/her research, but with a clear intention, there is an increased likelihood of predicting or noticing an impact.

PAR is the kind of research that gives life; in fact, I realized early that it is the kind of a life that gives life. In talk about sustaining life and the meaning of life, I also realized that these two aspects must be brought together because without meaning in life, there is no sustainability in life. When I first read about PAR, I was touched deeply because I had been discovering that engagement with people for/toward an objective that was beyond ourselves was fulfilling a personal need for spiritual communion or community with others. That was when I realized that that was one meaning in life, to live beyond my own satisfaction, beyond my own objectives and my own goals. I interpreted that to be one of the tenets of PAR, and I recognized it in my own work with others as empowerment, co-creation of new knowledge, transformation of self and environment. I saw that that is what the meaning in life is, that is what sustainability in life is. So PAR is not just about research, it is about life.

Choosing to engage in PAR is actually choosing to engage in a process where we know we will be transformed. We also know that what we do will transform others and will transform our mutual environment. So when we begin, it is with the knowledge that we will give ourselves up to this process. That is the source of the revolutionary nature of this transformation—it is revolutionary because we are actually transformed in the turning around experience created by PAR.

An Example of PAR

Between 1980 and 1987, I worked as a community-based educator with a provincial school jurisdiction legislated to serve twenty-five to twenty-eight Aboriginal communities in Northern Alberta. I was one of four persons appointed by the Ministry of Education to a commission established to provide educational support to the communities and to facilitate the process that would lead them to the governance of their own educational system. This was a historical movement in that those communities had never participated in the democratic election of representative school trustees and a school board. I found my thinking and my work entirely focused on cultural bridging between the knowledge systems of isolated traditional Cree communities and those represented by Alberta schooling systems. I had nothing in my formal training as an educator to prepare me for the kinds of expertise and knowledge that were needed, but my own background as a Métis person with roots in those Cree communities contributed immensely and immediately to my own capacity as a researcher and process facilitator, not only in depth and breadth of knowledge related to every aspect of the Cree world, but also in how that looked at the nexus of the school, the community, and the central administration. I did not know about PAR then but later years and more thinking have brought me to the place where I believe the principles of PAR most aptly describe those years and experiences. The learning and transformation of everyone involved in that process were beyond my expectations. The communities in the next few years initiated far-reaching changes in school policies and programming to promote school achievement for their children in ways that would not compromise their cultures and traditional knowledge systems. Despite the fact that almost all of the communities had Cree-speaking trustees, the local boards struggled against subtle but ongoing opposition to using their own language in the governance processes. Such opposition from predominantly non-Aboriginal administration was indicative of the general resistance from "standard" school practitioners to have a system that reflected Aboriginal forms of knowledge in any substantial way. This tension underlay every part of the schooling system and continues into today.

During one of my earliest trips to the communities, I received a warning from a wise old political leader, who said, "You believe in this structure and you believe in what you're doing, but we want to warn you, the system can't be changed unless you change the bureaucracy as well. You will change the governance structures but if there are no changes to the bureaucracy and administration, nothing will change." It was at least ten years before I realized the truth of that statement.

The PAR process in the communities was a time for integrating ancient knowledge into a new system of education for our people. I was not an outsider being true to a theoretical, albeit ethical, process. I was one of the people who wanted my children and grandchildren to be respected and cared for by teachers and administrators who cared about them as distinct beings representing ancient systems of knowledge and ways of being.

It was during the struggle to maintain the support to the community local school boards that I realized how accurate had been the predictions of my political advisers when we had first started on this long journey. Northland School Division continues to be dominated by non-Aboriginal administrators and staff, most of whom are unprepared professionally to provide leadership in dealing with the issues of Aboriginal education in any context.

What does this narrative have to do with PAR? During the time that I have described, the jurisdiction itself was being pressured to unfold in new ways. It was going through a living, vital process. As long as that was in place, the participation and the action of the people was sustained. Life and education was continuing. People were learning. People were involved. People were participating. There was anger, there was frustration but there was the excitement of learning. People wanted to be at the meetings, they wanted to participate. They saw hope because they were engaged. As long as these things were in place, PAR was happening. Life was happening. When some people wanted to keep the power, they started to try to control the movement. They put administrative procedures in place to ensure predictability and manageability. When that happened, the process stopped. There was no longer participation, there was no longer hope, there was no longer anything living and vital about the system. At that point, PAR no longer existed.

PAR Conclusion

I go out on the lake to pull the net, and that is PAR within an Indigenous knowledge paradigm. My father, who passed in 1988 to the spirit world, comes and has this conversation with me: "You have to bring the kids, my grandchildren. You have to bring them here so they can fish like this. They have to have the experience, they can't learn without the experience. You can talk about it but that is nothing, it doesn't mean a thing. They have to do it and they have to do it repeatedly." Repeatedly, because if you try to do it only once, it is not going to be part of who you are. In other words, you will not be transformed, and you will not be enhancing somebody else's experiences, which is what transformation is about.

Transformation is carried by intentions and commitment, and commitment includes emotions as well as intentions. Cocreation of knowledge is not merely an intellectual, empty sharing from which nothing grows. Creation means life, so knowledge, if it is to be life-giving and life-sustaining, has to have something that is based on life. It has to begin from life and give back to life, so it has to be connected to life in both the giving and the receiving. The only way that that can happen is with the full involvement of the whole person in experience. Your emotions and your heart are also to be engaged in PAR with the commitment coming from the heart.

To engage in PAR, a researcher has to trust, has to have faith, has to be able to say, "I want this" without knowing the end of the journey. An effective PAR researcher knows that the work means engaging and being truly in that state of communion between human beings who are both simultaneously the researcher and the researched.

As an Indigenous researcher, I can know that I want peace, I want change, I want my people to live, I want transformation, and I want knowledge about that transformation. But, more importantly, I am willing to work and give of myself totally for these things. I want that for any researcher who engages with Indigenous peoples and communities because that is what will open the heart to a turning around, a revolution, an inner freedom to reach out, fearlessly, and be transformed by giving of yourself. That is the spiritual base of the work of PAR, and that is the energy that lies at the interface of PAR with the ancient foundations of Indigenous research methodologies as these are lived out by Indigenous peoples.

References

Brown, L., & Strega, S. (Eds.). (2005). *Research as resistance: Critical, indigenous and anti-oppressive approaches*. Toronto: Canadian Scholars Press.

Ermine, W. (1995). "Aboriginal Epistemology." In M. Battiste, & J. Barman (Eds.), *First Nations education in Canada, the circle unfolds*. Vancouver: University of British Columbia Press.

Fals-Borda, O., & Rahman, M. A. (1991). *Action and knowledge Breaking the monopoly with participatory action-research*. New York: Apex Press.

Fals-Borda, O. (1997). Participatory action research in Colombia: Some personal feelings. In R. McTaggart (Ed.), *Participatory action research: International contexts and consequences*. Albany: State University of New York Press.

Freire, P. (1970). *Pedagogy of the Oppressed* (M. Bergman Ramos, Trans.). New York: Continuum.

Gott, R. (2008, August 26). *Orlando Fals Borda: Sociologist and activist who defined peasant politics in Colombia*. The Guardian, UK. Retrieved from http://www.guardian.co.uk/world/2008/aug/26/colombia.sociology

Hampton, E. (1995). Towards a redefinition of Indian education. In M. Battiste, & J. Barman (Eds.), *First Nations education in Canada, the circle unfolds*. Vancouver: University of British Columbia Press.

Meyer, M. (2002). Acultural assumptions of empiricism: A native Hawaiian critique. *Canadian Journal of Native Education, 25*(2), 188–198.

Steinhauer, E. (2008). *Parental school choice in first nations communities*. Saarbrucken, Germany: VDM Verlag Dr. Mueller.

Weber-Pillwax, C. (2003). *Identity formation and consciousness with reference to Northern Alberta Cree and Métis Indigenous peoples*. Unpublished Doctoral Dissertation, University of Alberta, Edmonton, Canada.

CHAPTER FIVE

Ko tātou te rangahau, ko te rangahau, ko tātou*: *A Māori Approach to Participatory Action Research*

LYNNE HARATA TE AIKA &
JANINKA GREENWOOD

Ehara toku toa he toa takitahi, engari he toa takitini
My strength is not of one, but of many

Introductory

Participatory Action Research is an approach to research that not only integrates action and investigation, but also involves those who would be most affected by the project as co-researchers. Although in Western scholarly traditions it is a relatively recent, though widely adopted research paradigm (Kemmis & McTaggart, 2005), in Māori traditions it is embedded. It is embedded in the workings of the *marae*, the people's communal house, where the community investigates and debates issues that concern its well-being and future actions. Cycles of action, reflection, and reconceptualization are involved. *Tātou ki a tātou* (All of us together are working for all of us) is a phrase that repeatedly reoccurs. This phrase is embedded in mythic understandings of knowledge: Tane climbed to the heavens and brought back and shared three *kete* (baskets of knowledge). Each provided access to a particular sphere of knowledge from the esoteric to the practical, and each was assigned to those who would best explore and develop that knowledge for the well-being of the people. The phrase is also embedded in contemporary *kaupapa* Māori research, which upholds the building of capacity and the development of well-being for the *iwi*, or people, as the primary goals of research, which emphasizes the importance of accountability to *iwi* and signals preference for methodologies and methods that engage participants as co-constructors of the findings.

Focus of This Chapter, Who the Writers Are, and Some Initial Terms

This chapter develops a conceptual framework that loosely aligns Māori approaches to knowledge, research, and action with Western theorizations of participatory action research. It offers illustrative examples taken from a number of projects, including current ones such as *Kotahi Mano Kāika*, Ngāi Tahu's strategy for language revitalization, and *Hōaka Pounamu*, a University of Canterbury College of Education Māori language immersion program for practicing Māori teachers, and a historic project, *Te Mauri Pakeaka*, a program for developing cross-cultural understanding through the arts. It also suggests ways, again with illustrations from specific projects, in which Māori conceptualizations can enrich understandings of participatory action research.

We, the writers, come from Māori and immigrant perspectives respectively. Lynne Harata is Ngāi Tahu, Ngāti Awa, and Te Whānau Apanui. Born in the South Island, which has a relatively low density of Māori, she has devoted much of her working life to language revitalization and the development of educational opportunities for Māori by Māori.

Janinka is a Czech-born immigrant to New Zealand. The teachers in her adopted country have been both Māori and European. Her research pathways, and consequently her teaching, have been continuously engaged with the exploration of cross-cultural perspectives and cross-cultural approaches to knowledge.

We have been close colleagues for the past eight years, working together on a range of strategic educational developments and research projects, some of which are described in this discussion. Although in the discussions that lead to this writing we draw on our particular individual perspectives, here we offer the distillation of that discussion and our ongoing investigation.

Our discussion in this chapter is specifically located in New Zealand, and in the indigenous and the cross-cultural concerns that are particular to New Zealand. However, we think that many of the concerns and conceptualizations we discuss will have significance for other communities around the world. Consequently we use terminology and metaphors that would for the most part be readily understood within the New Zealand context but we offer an explanation of some key terms here at the start of our discussion. Others we will explain as they arise.

As many of our readers will know, Māori are the indigenous people of Aotearoa New Zealand. *Pākehā* is the term most commonly used to refer to non-Māori immigrants of British and European descent. Sometimes in media and popular discussion, the word *Pākehā* is used as a more general term marking the mainstream. For example, the *Pākehā* education system might refer to all of New Zealand's education system excluding Māori initiatives, and *Pākehā* values, might refer to the value systems of the Western world.

We use Māori *whakataukī*, or ancestral proverbs, where they seem appropriate. The Māori tradition of knowledge and debate is an oral one, and within that oral tradition *whakataukī* serve as reference points that anchor lines of thought to recognizable traditions, very much in the same way as citations anchor scholarly articles in a Western tradition. For example the title of our chapter, *Ko tātou te rangahau, ko te rangahau, ko tātou,* references a phrase often used to proclaim lineage and regional identity: *ko au, ko te awa, ko te awa ko au,* claiming the interdependence of the speaker and the ancestral river. Here we have adapted the *whakataukī* to emphasize the alignment between the action of research and the people affected as both participants and beneficiaries of the research.

In looking for a metaphor for our discussion, we turned to the image of *whare kōrero*. Literally this might translate as a house of speech, but more particularly it is a term we have used in our joint research projects to describe a discussion site, either online or live, in which participants collaboratively explore findings and co-construct emergent narratives. It is a house not only built through talk but also constructed to safely house and support further talk. Just as a physical *whare* would have carved *pou* or posts to mark its history and the lineage of the people it belongs to, so our *whare kōrero* has four *pou* that are the projects we describe. A physical *whare* might have its floor covered by a *whāriki,* or finely woven mat, and so our *whare kōrero* has a *whāriki* of discussion in which we tease out the various threads of our understanding of Māori approaches to action research.

In the sections that follow we describe each of the *pou* of our discussion, each of the case studies, separately. Then we weave the *whāriki,* the pattern of concerns, approaches, and themes that grow out of our review of the individual projects and that offer a grounding for further exploration and elaboration.

Kotahi Mano Kāika: Ngāi Tahu's Strategy for Language Revitalization

The first *pou* in our *whare* is a project that Lynne Harata has been deeply involved in, but that extends beyond well her. It is Ngāi Tahu's project of language revitalization. Ngāi Tahu is the *iwi* or tribe who asserts chiefly rights to most of the South Island of New Zealand. The history of alienation of its lands in the first decades of colonialism led to its sociopolitical marginalization and consequently to the loss of language. The details of its treaty claims and the establishment of a corporate business plan are outside the scope of this chapter.[1] What is relevant is that during the 1990s Ngāi Tahu began a program of language revitalization. In the previous decade, Benton (1981) had predicted that the Māori language was in the process of dying. The 1970s, 1980s, and 1990s saw programs of language revitalization throughout the North Island, particularly among tribal groups with significant numbers of native speakers.

Ngāi Tahu's project took place in the second wave of revitalization. Ngāi Tahu's strategic plan (Ngāi Tahu, 2001) identified that the intergenerational transfer of Māori language as the main form of communication had not occurred within South Island Ngāi Tahu for approximately 80 years and in some areas for 130 years. It also identified that there were approximately three remaining native speakers still living within the tribe, and that while there were some *Kōhanga Reo* (early childhood centres with Māori immersion) and *Kura Kaupapa Māori* (schools operating in the Māori language only and with Māori value systems) in the South Island, these were insufficient to revitalize the language and to ensure intergenerational transfer. There were, however, people who affiliated with Ngāi Tahu living in the North Island who were fluent in the language, in many cases as a result of language initiatives by North Island tribes. The challenge was to build areas within the South Island with a high enough density of fluent speakers of Māori for intergenerational transfer to occur. It was calculated previously (Spolsky, 1987) that a thousand Māori medium teachers would be needed to enable a program of language revitalization. Twenty years later, in 2000, Ngāi Tahu on the advice of Fishman, a world recognized expert in endangered languages, began its language strategy by focusing on the home and intergenerational transfer. *Kotahi Mano Kāika, Kotahi Mano Wawata* (a thousand homes [where the language is alive and used] achieving a thousand aspirations) was the result.

The project is still in progress. Over the last decade, the cycles of development of the project have involved the overall strategic plan (Ngāi Tahu, 2001); an operational education plan, *Te Kete o Aoraki* (Ngāi Tahu, 2002), which not only placed access to learning of the Māori language among the reportable goals for all schools in the region, but also arranged for direct reporting to Ngāi Tahu on the educational outcomes of all Māori students; a Memorandum of Understanding with the Ministry of Education (Ministry of Education & Ngāi Tahu, 2002); and a plan for supporting the development of bilingual early childhood centres and *iwi* endorsed schools (Ngāi Tahu, 2005). In this context, a scheme was developed to establish an intensive language immersion program for Māori teachers (Te Aika & Greenwood, 2007), an in-school program for teachers building dialect and language capacity (Te Aika, 2008), the development of learning resources, a Web site, and an e-mail and text-messaging promotion strategy.[2]

Hōaka Pounamu: Immersion Program for Teachers

The second *pou* in our *whare kōrero* is a project that Lynne has initiated and led. Hōaka Pounamu is a year-long Māori medium program for teachers working in *kura kaupapa* Māori, teaching Māori medium programs within mainstream schools, or providing bilingual support and leadership within a mainstream school.

As a Ngāi Tahu initiative, it sits within the tribal goal of language revitalization. It involves a partnership of Ngāi Tahu, the Ministry of Education who funds the delivery and the teachers' release from school, and the College of Education within the University of Canterbury where the qualification is accredited and the program is delivered.

The program aims to build sufficient numbers of bilingual teachers in the South Island to parallel the development of Māori as a communicative language in the home, with enough strength and attractiveness in its usage to ensure intergenerational transfer. Specifically the goal is to graduate teachers with oral confidence as well as proficiency in the language, with a sound knowledge of *iwi* history and current aspirations, and with a strong grasp of second language pedagogies.

The name of the program, Hōaka Pounamu, references the sandstone (*hōaka*) rubbing against the greenstone (*pounamu*) to remove its rough exterior. The *pounamu* represents the language, a treasure handed down from ancestors and a connection with identity and place in the universe.

The evolution of the program might be described in terms of a *koru*, a spiral form that is derived from the unfurling frond of a fern. As the frond grows, series of new leaves bud, develop, and unfold. The *mauri* (life force) of the fern is there from the beginning, but each stage generates further growth. In this case the first leaves symbolize consultation with the *iwi*, ensuring not only that the direction and content were in keeping with *iwi* aspirations, but also that leaders and teachers within the *iwi* would come in to the program at various times to share their knowledge. Over the seven years that the program has run, it has been in a process of ongoing development using feedback from *iwi*, teaching staff, students, and schools.

A key feature of the program is that it places language as an integral part of *mātauranga Māori* (Māori knowledge). Each year this is highlighted by a week spent on one of the *marae* (tribal community ground or site of ceremony, usually with a meeting house as part of the complex) within the region researching the history, ecology, and spirituality of place, and developing teaching resources from this research to share with the people of the *marae*. The assignment fosters reciprocity of benefits to the host community and the students, building the capacity of all the participants to carry out their responsibility of *kaitiakitanga* (stewardship) of the land—in all its layers.

Developing Bicultural Understandings and Practice in the College of Education

The third *pou* in our *whare* is a project that both Lynne Harata and Janinka, together with other colleagues, have been involved with over a number of years. Initially the Christchurch College of Education was a separate tertiary institution, whose core business was initial and in-service teacher education. At the start of 2007 the college merged with the University of Canterbury, and it continues with its role of teacher education.

In 2001, the college could be described as a relatively monocultural establishment. In this way it reflected its surrounding city and most of the South Island. If change was to occur within schools, it needed to also occur within in the institution that prepares teachers. Lynne Harata had recently been appointed as *Kaiwhakahaere Māori,* a leadership position in the school's management team, and Janinka and Liz Brown were seconded as coordinators of the bicultural project (Greenwood & Brown, 2007).

The first step was to develop a strategic plan for the college, following internal and external consultation and to negotiate its adoption by the strategic leadership team. The plan drew on the government's goals (Ministry of Education, 2003) for Māori development and education and on the Memorandum of Understanding between Ngāi Tahu and the Ministry of Education.

One of the first agreed action points, and an example of our approach, was to develop and deliver a course that would introduce students to the Treaty of Waitangi and its relevance to the present day and to their work as teachers. The challenge was to develop a course that was not didactic, yet provided access to significant new knowledge and allowed the participants to be active in the process of investigation, exploration, and meaning making. The course that evolved involved a collation of resources, providing an initial platform for participants' research and a series of tasks asking them to use one or more of the arts to present their findings. The choice of art as a medium for analysis and presentation was a deliberate move away from a positivist approach to one that would allow participants to own their questions and their learning. "Boss of our story" was the title of an article in which we reported on progress (Greenwood & Brown, 2004). The course delivery and development continues with similar goals but in the hands of new facilitators.

Further strategic goals included the development of a praxis of Māori language and cultural practices within the institution as a whole, the development of Māori and bicultural research projects, and the re-evolution of all curriculum areas to include Māori perspectives.

The business of merger between the college and the university temporarily slowed down and complicated progress in the pursuit of these goals. However, the building of staff and student capacity in these areas continues as an ongoing project.

Te Mauri Pakeaka

The fourth *pou* of this *whare* takes us beyond the region of Ngāi Tahu and to a decade or so earlier to a project led by Arnold Wilson in which Janinka was a participant. *Te Mauri Pakeaka* (Greenwood & Wilson, 2006) was a project set up within the education system in the mid-1970s with the aim of changing the face of mainstream society from one that was unconsciously monocultural to one that recognized and celebrated its Māori population and heritage. Over the fifteen years that the project ran, it brought clusters of teachers, students, artists, and elders to three- or five-day workshops on *marae* throughout the country.

The immediate objective in each workshop was to create an art work, visual or performance-based on a local history. However, the process of living together on the *marae,* and in that context researching the history for their work, allowed them to experientially explore what a significantly bicultural society might be like. Elders who had previously been shut out from the education system, found ways of teaching not only their young people, but also teachers and departmental advisers. *Pākehā* students and teachers learned about Māori values through living in an environment where they were the norm, turning to their Māori colleagues for leadership. Māori students and teachers took that leadership in a range of ways and learned how their traditional beliefs and ways of relating could have currency within their education.

Arnold Wilson repeatedly talked about art being a catalyst (Greenwood & Wilson, 2006; Wilson & Greenwood, 2004). In its first years, it was a catalyst to recognizing, exploring, and appreciating difference, for discovering the human, social, historical, and spiritual dimensions of place, and for enabling people to vision what a bicultural education system and society might be like. By the 1980s, the sociocultural environment in New Zealand had changed: Māori assertion was stronger, Māori initiatives for change were pronounced, and institutional systems, such as education, were under overt scrutiny. The process of art making within the *Pakeaka* workshops was now a catalyst for participants to find their ground within the change. Capacity was being built for Māori participants to more fully understand their heritage and to explore the leadership roles they might take in change, for *Pākehā* to reconfigure their assumptions, come to terms with change and explore their own roles, and for partnership to develop. The supportive environment and the use of art as a medium for exploration created a safe space where participants could explore ideas and roles which initially might have felt unsafe.

The Whāriki: Patterns and Themes

Tino Rangatiratanga

Each of the projects we have described has dual goals of *tino rangatiratanga* (development of the capacity to be self-determining) and *kaitiakitanga* (the stewardship of resources). Ngāi Tahu's language revitalization project explicitly looks forward to the year 2025 as a period by which it aims to have an operational tribal ecology in place. Such an ecology has a vital language, land and water resources, and a people in balance with their history, language, and environment. The *Hōaka Pounamu* program also explicitly aims to develop teachers with the knowledge and skills needed to make them active contributors to the tribal goals of self-determination, stewardship, and the intergenerational transfer of the language. *Ko te reo te pūtake o te Māoritanga* (language is the deep root which nourishes the growth of a Māori way of being in the world). The college bicultural project had the same broad goals, particularly with the treaty course, focusing on the importance of ensuring that graduating teachers would be informed and willing participants in the connected project of societal change.

The goal of *tino rangatiratanga* is also at the base of *Te Mauri Pakeaka*, though in a more complex way. On the one hand, the project was directly geared to reempower Māori and guide the education of their young people, in terms of content, resources, and processes. On the other hand, it was also geared to enable *Pākehā* to find their own role in supporting Māori self-determination and exploring their identity within the changed society that would evolve through Māori self-determination.

Although at a time of disempowerment self-determination is often seen as a goal, it is a process: there is something that the people are self-determining about. The overarching goal of self-determination in the projects we describe is one of well-being for the people. Different tribal communities express it in various ways. Ngāi Tahu sums it up in this mantra: *Mō tātou, ā, mō ka uri ā muri ake nei* (for us all and for the generations to come).

Participation, Ownership, and a Range of Levels of Entry

The relationship between project goals and participants is important in each of the four projects. There is an underlying understanding in each that knowledge is most useful when it is accessed and developed by those who need to use it.

Ngāi Tahu's language revitalization project depends on engagement of the people as a whole in its aims and its strategies. Tribal consultation *hui* (meetings) took place in the late 1990s. At this time the tribe's treaty claims, which had engaged the energy of three generations, were being settled and Ngāi Tahu was in a position to reinvest its treaty settlement money and energy into tribal development. The consultative model is one that already exists historically on the *marae*, where issues are presented and debated for the people to find answers together. It was through these consultative *hui* that language, culture, and identity were prioritized as key areas for development. Each of the initiatives within the overarching project, such as the launching of *Te Kete o Aoraki* and the text message campaign, was a means of bringing more participants into the work of development and of encouraging them to take ownership of the process and the outcomes.

Project participants need to find a level of entry that is both comfortable and meaningful for them to feel fully engaged. One of the deterrents to language acquisition is a perceived level of difficulty that takes participants too far outside their comfort zone. So the *Hōaka Pounamu* program addresses this by staircasing learning in a succession of courses that progress from bilingualism to total immersion, and by building on the *tuakana-teina* approach. In family terms the relationship of *tuakana* and *teina* is that of older and younger siblings. In the context of investigation and learning, *tuakana-teina* expresses the notion that people come with different skills and areas of expertise and so alternatively take the role of *tuakana* or *teina* in a range of activities, the more skilled helping the beginners. In this way the teacher may well be a *teina* in some areas of knowledge in relation to one or more of the students in the class. This is a concept that is crystallized in the word *ako,* which refers to both teaching and learning, and that as a current educational concept describes the process of collaborative learning through teaching.

The notions of *ako, tuakana-teina,* and of staircased entry also characterize the other two projects we represent as *pou* in our *whare*. The widely mixed range of participants in the *Pakeaka* project meant that different groups and individuals brought different degrees and kinds of knowledge. Being supported in finding their own place to begin their investigation and representation allowed each of them to take ownership of the goals and to process and foster a sense of communal ownership of all the exploration and discoveries that took place in the workshop. The same approach was used on a smaller scale in the treaty courses. Participants came with different knowledge, assumptions, and attitudes. The course recognized that there was no absolute or externally situated knowledge that would be useful to them. It was only knowledge they owned that would make a difference and the discovery of that knowledge had to take place at their own level of entry. However, the group as a whole could be *tuakana* and *teina* for each other at all of those levels.

The concept of *ahi kā* is important in the Māori world. It refers to the need to keep the home fire burning and warm to maintain the right to occupancy. Knowledge that is not owned might be conceptualized as an *ahi matao,* a fire that has gone out and is cold. Participant ownership is a fire that is warm, an *ahi kā.*

Nested Projects, Cycles, and Bricolage

Clearly the first three *pou* that we place in our *whare* present a series of nested projects. While each might in a different context have occurred independently, each actively draws on its connections with the others. At the same time, each of the *pou* can be seen as a composite of smaller separate projects. The fourth *pou,* a succession of workshops, can also be seen as a series of nested projects all building to the same goal.

The *whakataukī* (ancestral proverb), *nāu te rourou, nāku te rourou, ka ora ai te manuhiri,* addresses the importance of collaborative endeavour. Literally translated it says, with my food basket and your food basket the multitudes can be fed. At the level of metaphor it talks about the different talents and energies participants can bring to a project and the value of taking into account and utilizing each of those different, and perhaps unexpected, talents and energies.

The Western concept of bricolage (Denzin & Lincoln, 2000) positions the researcher as a bricoleur, one who uses a mixture of investigative strategies as needed, who identifies gaps in the data or the project development and looks for further strategies to adopt. Such a concept acknowledges the significance for some projects of evolving design and of considering multiple dimensions of problems (Kemmis & McTaggart, 2005). In the projects we have described, the concept of bricolage might be conceptualized as a communal approach to investigation. The project leaders bring their visions and skills, but leave room for all the participants to add their contribution and influence the direction and outcomes. In the *Hōaka Pounamu* project, the shape of the work changed according to the connections and skills of each group of participants. In some years, because of the knowledge of particular participants the relationships between an art form, language, history, and identity were explored; in another year, it was the depth of history and oral traditions of a particular *marae*. In a program that was the first, and that remains the only Māori language immersion program for teachers in the South Island, such a layering of expertise develops the platform of knowledge from which future delivery can be built. It also develops *ihi* (charged energy) of the program in such a way that it can inspire and sustain participants. In *Te Mauri Pakeaka,* the layering of expertise was an essential ingredient of the project. The design acknowledged that the monoculturalism within schooling and the poverty of outcomes for Māori students as a whole could not be solved by the actions of one group: elders, teachers, students, families, administrators, and artists had to come together, and, most importantly, acknowledge that each of the other groups held valuable knowledge and skills that were needed in the search for resolutions.

Earlier we suggested the unfurling fern frond as a metaphor for the way each of these projects evolves. The unfurling suggests a sense of organic development: a development that is both planned and unplanned at the same time, or that is planned in a way that invites and uses further and even unexpected contributions and evolutions. It also suggests that growth is subject to interruption and breakage, but that these do not prevent its continuance. *Mate atu he tetekura, haere mai he tetekura,* the *whakataukī* tells us: when one tree frond dies down another rises to take its place. The overall goal of capacity building and *tino rangatiratanga* give form to a new project, scaffolding on what has already been achieved. *Ko te pae tawhiti whāia kia tata, ko te pae tata whakamaua kia tina* (seek out the distant horizon, but work with what is close at hand).

Reciprocity and Succession Planning

The development of leaves on each side of the unfurling stem is also suggestive of the concept of reciprocity that is inherent in the projects. As *Te Mauri Pakeaka* illustrates, there are gains for all parties involved. In this case, as explained before, teachers and administrators gain not only the opportunity to better understand the culture of their students, but also to see their students achieve in ways they had not expected, and to engage the Māori community in the process of schooling. Māori elders, while giving their knowledge, gain not only a recognition of the value of that knowledge within the system of schooling, but have the opportunity to reinscribe the system so that their negative experiences of schooling do not have to be the same for their children and grandchildren. And so on for each of the groups of participants.

Partnerships are a pivotal factor in the design of each of these projects. The participants are construed as partners within the project not as consumers or objects of investigations. In addition, strategic partnerships with those who hold, by using or blocking, power within the mainstream are also important. Ngāi Tahu worked with international language theorists to develop its strategic plan. It selected the College of Education as the agent to deliver its immersion project for teachers, rather than developing its own autonomous institution. On a broader front, all of the projects, although their goal is capacity building for Māori recognize that, at some level or another, partnership with *Pākehā* is important. If *kaupapa* Māori is to be fully developed and operational at a national level, there needs to be opportunity for non-Māori to find their place in it or alongside it—or it will always be in contestation and conflict.

Implicit in this approach is recognition of the importance of succession planning. Each of the projects we have described acknowledges that the researchers need to develop those who will take over the further cycles of the project.

Agents of Change

Although a Māori perspective constantly draws on traditional knowledge, it also clearly recognizes that the challenge is to live in *te ao hurihuri* (the contemporary world of constant change). The goal of capacity building is thus not a static one; it is one of preparing for constant and complex change. *Te manu kai i te miro, nōna te ngahere; te manu kai i te mātauranga, nōna te ao.* This *whakataukī* tells us that the bird that eats the berry from the miro tree can be at home in the forest; the one that eats knowledge can be at home in the world. The goals of capacity building allow participants to access the kinds of knowing and knowledge that will enable them to manage the evolving world. Each of the projects we have described is designed to empower its participants to be agents of change. Karetu, honorary father of Māori language revitalization, identifies language as a sign of prestige and potency: *tōku reo, tōku ohooho; tōku reo, tōku mapihi maurea; tōku reo, tōku whakakai marihi* (language is signal of and tool for identity, chiefly status, and ability to act potently in the world).

Complexity theory (Davis, Sumara, & Luce-Kapler, 2008) positions knowledge as a complex mesh of interactions that reach from neurology through psychology, sociology, anthropology to ecology, and learning as transphenomenal, transdisciplinary, and interdiscursive. Such a construct parallels the concept of *ao hurihuri* and the need for knowledges that are complex and evolving.

Holistic Understandings

The projects we have described engage participants in terms of *wairua* (spirituality), *ngākau* (emotions), *tinana* (physicality), *whānau* (family) as well as *hinengaro* (intellect). Knowledge and knowing may be conceptualized in all these dimensions, and so may the development of capacity.

Investigation in each of these projects, therefore, takes place not only in terms of data as it might be understood within a Western construct but also in terms of identity construction, *kaitiakitanga, rangatiratanga,* and *mana* (esteem or value). In other words investigation is not really separable from its products. Knowledge is only meaningful when people own it and use it, spiritually or in action. In this way there is no cut and dry differentiation between the concept of research and that of learning.

Kaupapa Māori and Participatory Action Research

As we teased out the strands earlier, we have suggested some parallels to Western models of participatory action research. We have not attempted a detailed paralleling.

Smith (2005) locates the arena of indigenous research in terms of a desire to avoid dominant culture definitions. "The desire of 'pure', uncontaminated, and simple definitions of the native by the settler," she writes (Smith, 2005, p. 86), "is often a desire to continue to know and define the Other, whereas the desire by the native to be self-defining and self-naming can be read as a desire to be free, to escape definition, to be complicated, to develop and change, and to be regarded as fully human." She points out that indigenous communities are not homogenous, and therefore do not agree on all issues and seldom live in isolation from the rest of the world. However, she also notes that there are still many indigenous communities "who possess the ancient memories of another way of knowing that informs many of their contemporary practices" (Smith, 2005, p. 87).

Accordingly, the research methods used in *kaupapa* Māori research may range from a variety of quantitative data gathering processes to ones involving action, art products, or narratives. What characterizes *kaupapa* Māori research is a concern with capacity development, utilization of Māori ways of knowing, participation, collaborative ownership, and respect for *mana*.

Bishop (2005) discusses how educational researchers have often been slow to acknowledge cultural difference, and how "as a result, key research issues such as power relations, initiation, benefits, representation, legitimization, and accountability continue to be addressed in terms of the researchers' own cultural agendas, concerns, and interests" (p. 110). He suggests "such domination can be addressed by both Māori and non-Māori educational researchers through their conscious participation within the cultural aspirations, preferences, and practices of the research participants" (p. 110).

As our *pou* illustrate, the most significant feature of Māori research is that its goals originate with Māori and are consistent with Māori values. Research is branded as something bigger than academic knowledge.

Notes

Literally, we are the research and the research is us.

1. Readers who want to know more might like to consult Evison (1993) Te Waipounamu The Greenstone Island; and www.ngaitahu.iwi.nz
2. Readers are invited to visit these Web sites: www.generationreo.com and www.kmk.Māori.nz

References

Benton, R. (1981). *The flight of the amokura: Oceanic language sand formal education in the South Pacific*. Wellington: NZCER.

Bishop, R. (2005). Freeing ourselves from neo-colonial domination in research: A kaupapa Māori approach to creating knowledge. In N. Denzin, & Y. Lincoln (Eds.), *The Sage handbook of qualitative research* (3rd ed.) (pp. 109–138). Thousand Oaks, CA: Sage.

Davis, B., Sumara, D., & Luce-Kapler, R. (2008). *Engaging minds: Changing teaching in complex times*. New York: Taylor and Francis.

Denzin, N., & Lincoln, Y. (2000). Introduction: The discipline and practice of qualitative research. In N. Denzin, & Y. Lincoln (Eds.), *Handbook of qualitative research* (2nd ed.) (pp. 1–32). Thousand Oaks, CA: Sage.

Evison, H. (1993). *Te Wai Pounamu The Greenstone Island: A history of the southern Māori during European colonisation of New Zealand*. Christchurch: Aoraki Press.

Greenwood, J., & Brown, L. (2004). Boss of our story. *New Zealand Journal of Teachers Work*. Online Journal, 1(2).

———. (2007). The treaty, the institution and the chalkface. In T. Townsend, & R. Bates, (Eds.). *Handbook of teacher education: Globalisation, standards and professionalism in times of change* (pp. 67–78). Dordrecht: Springer-Kluwer.

Greenwood, J., & Wilson, A. (2006). *Te Mauri Pakeaka: A journey into the third space*. Auckland: Auckland University Press.

Kemmis, S., & McTaggart, R. (2005). Participatory action research: Communicative action and the public sphere. In N. Denzin, & Y. Lincoln (Eds.), *The Sage handbook of qualitative research* (3rd ed.) (pp. 559–604). Thousand Oaks, CA: Sage.

Ministry of Education. (2003). *Education priorities for New Zealand 2003*. Wellington: Ministry of Education.

Ministry of Education & Ngāi Tahu. (2002). *Memorandum of understanding*. Christchurch: Ministry of Education.

Ngāi Tahu. (2001). *Kotahi mano kāika*. Christchurch: Ngāi Tahu Development Corporation.

———. (2002). *Te kete o Aoraki*. Christchurch: Ngāi Tahu Development Corporation.

———. (2005). *Kotahi mano kāika 5-year implementation plan*. Christchurch: Te Rūnanga o Ngāi Tahu.

Smith, L. (2005). On tricky ground: Researching the native in the age of uncertainty. In N. Denzin, & Y. Lincoln (Eds.), *The Sage handbook of qualitative research* (3rd ed.) (pp. 85–108). Thousand Oaks, CA: Sage.

Spolsky, B. (1987). *Report on Māori-English bilingual education*. Wellington: Department of Education.

Te Aika, L-H. (2008). *Milestone 1 Report to Ngāi Tahu, Reo Kura*. Unpublished report to Te Rūnanga o Ngāi Tahu.

Te Aika, L-H., & Greenwood, J. (2007). *Hei Tauira: A case study of Hōaka Pounamu a bilingual teacher development program*. Paper presented at Language, Education, and Diversity Conference, November 23, 2007. Waikato, New Zealand.

Wilson, A., & Greenwood, J. (2004). Shifting the centre. *Perspectives in Education, 22*(1), 151–149.

CHAPTER SIX

Translating "Participation" from North to South: A Case Against Intellectual Imperialism in Social Science Research

Cynthia M. Chambers &
Helen Balanoff

> ...all societies, particularly the "traditional" ones, have been, and remain, participant. None of them needs to be trained or initiated by outsiders or professionals to the "secret" formulas of modern participation.
>
> —Rahnema (1990)

Participatory Research in Northern Indigenous Communities

Participatory action research (PAR) aims for social justice in both the process and the outcome of research. To achieve this aim, PAR must contain and support the social and material conditions necessary for meaningful participation in a research project, particularly by those subordinated within asymmetrical relations of power. What makes PAR participatory, and thus socially just, is contingent on opportunities provided for those involved to exercise their capacities, to express their experiences, and to participate in determining their actions and the conditions of those actions (Young, 1990). The dichotomy between first peoples and settlers (Sissons, 2006) complicates participation in PAR research in northern Canada. As part of its social justice agenda, PAR aims to have local people, that is indigenous peoples in the North, become the experts and agents of the research agenda, rather than being the other and the object of study, as they are often positioned in conventional social science research (Absolon & Willett, 2004). With PAR indigenous knowledges and knowledge practices, as well as indigenous languages and voices, become indispensable to the research (Sinclair, 2003). PAR researchers, in the North and elsewhere, aim to balance attention to the social processes of the research with social outcomes of the research. This means that researchers must not universalize Western processes of knowledge formation, inherited from European intellectual traditions, to northern indigenous communities. For northern indigenous communities, all animate beings, which include the human and the nonhuman as well as the ancestors, generate knowledge within the "spheres of nurture" where they dwell. In institutionalized social science research, professionals generate knowledge, which is then assumed to be either universal (i.e., theoretical) or limited to a specific population rather than a location (i.e., empirical).

PAR assumes mutuality between outside researchers and the local communities with whom they work, an arrangement where both have capacity and opportunity to participate fully (Heshusius, 1994), particularly in decision making (Bishop, 1998; Cornwall & Jewkes, 1995). For Young (1990), this mutuality is the basic condition for social justice. In the Canadian North, the aim is for more than mutuality; most northern indigenous communities want ownership of the research. This means that local communities initiate and set the agenda; accrue benefits from the research; represent local knowledge using indigenous language, voice, and media; ensure the research is authorized and validated locally; and hold the researchers, local and otherwise, accountable to the communities whom the research serves (Bishop, 1997). The question of ownership over research is especially salient in northern Canada, the colony within (Watkins, 1977), where the state assumed sovereignty over northern indigenous peoples and the lands they inhabit. One legacy of colonialism is that northern residents are positioned dichotomously as either indigenous or not, on either side of disputes about title to land and ownership of resources. Communities in the North originated as locations of trade with, and evangelizing of, indigenous people. Eventually these trading posts and Christian missions became sites of colonial settlement and resettlement, "contact zones...social spaces where disparate cultures meet, clash and grapple with each other, often in highly asymmetrical relations of domination and subordination" (Pratt, 1992, p. 4). Within these northern contact zones, organic social processes, such as ways of participating in community life, have grown up, and these are inseparable from the indigenous epistemologies to which they are akin (Absolon & Willett, 2004). Community persons bring these local ways of being, acting, and communicating to the social processes of a research project, and then those projects become another set of intersecting and interconnecting social relations within a community. Many northern researchers have used participatory research as a framework to facilitate more symmetrical relations of power between themselves and the indigenous communities with whom they work. With the participatory approach, local communities can control decision making and verify results (Kritsch, 2002; Legat, 1995; Rabesca, Romie, Blackduck, Zoe, Legat, Johnson, & Ryan, 1993; Ryan, 1994, 1995; Thorpe, Hakongak, Eyegetok, & Kitikmeot Elders,

2001). These projects thus foster socially just and locally appropriate participation, not as defined elsewhere but as practiced and understood within the context of particular contact zones.

In the following section, we introduce the Ulukhaktok Literacies Research Project (ULRP) and describe and analyze efforts to ensure the project was participatory within the context of a specific northern community.

The Ulukhaktok Literacies Research Project

Ulukhaktok[1], a community of four hundred in the Northwest Territories (NWT), Canada, is the site for a multiyear research project. The initial research question was: what constitutes literacy for *Ulukhaktokmiut* (the people of Ulukhaktok)? We were interested in literacy before the introduction print with colonization, as well as contemporary literacies. The study was designed to focus on the local language, knowledge, experience, and expertise of the Ulukhaktokmiut.

A Colonial Context

Northern communities were "'made' rather than 'born'" (Madan cited in Cornwall & Jewkes, 1995, p. 1673) and Ulukhaktok was no exception. In the 1920s, Ulukhaktok was a location for trade on Victoria Island in the Canadian Arctic. By the 1930s, Inuit from Alaska and the Western Arctic, who had migrated to the region, moved into the tiny mission and trading-post settlement. The following two decades saw *Inuinnait*,[2] particularly the *Kangiryuarmiut,* move into Ulukhaktok, as well (Collignon, 2006; Stern, 1999). In the 1960s, most people still made their living on the land, hunting, fishing, and trapping for furs. However, the people now lived year round in Ulukhaktok rather than on the sea ice in winter and on land from spring through autumn. Different Inuit groups, who traditionally inhabited, and travelled extensively around vast territories from the extreme west of Alaska to the Central Canadian Arctic, were now settled permanently together in small isolated arctic communities like Ulukhaktok. In these communities, the different Inuit groups also now lived alongside *Qablunaat* (the White people or non-Inuit Westerners), who were there to minister, administer, or trade.

A Site of Métissage

Communities, like Ulukhaktok, which grew up in colonial terrain became sites of métissage (Chambers, Hasebe-Ludt, Leggo, Oberg, Donald, & Hurren, 2007; Lionnet, 1989), places where histories and discourses mixed, and where a vernacular of the colonial language became the local *lingua franca*. In the early years, it was more common than it is today for Ulukhaktokmiut, including Qablunaat, to speak one of the indigenous languages. By the 1960s and 1970s, however, English had overtaken the indigenous languages. Today most people speak a local vernacular of English, although *Inuinnaqtun* and *Inuvialuktun,* the dominant indigenous languages are still audible in the community.

Although their geographic origins may differ, and their languages may vary, contemporary Ulukahktokmiut believe that they hold in common a way of life and a shared set of values and beliefs. Their desire that young people not forget *pitquhiraluavut* (our ways), that is, these shared values and beliefs, and way of life, fuelled community interest in the ULRP.

Research Agenda

Since the introduction of English literacy in the North, "alphabetized communication is expanding at the expense of all other forms of pattern-languaged understanding" (Rasmussen, 2000, p. 14). Pattern-languaged understanding is communicated through skilled practices (Ingold, 2000) such as dancing, singing, drawing, and sewing (Rasmussen, 2000). With alphabetic literacy came scriptist-bias and "alphabetization of thought" (Rasmussen, 2000), where knowledge and knowledge practices are reduced to discrete, decipherable items amenable to codification and representation schematically (Ingold, 2000) for consumption as, for example, school curriculum or public policy. When communication and knowledge practices are alphabetized, literacy becomes singularly linguistic to the exclusion of pattern-languaged communicative practices such as sewing and drawing, skills at which Ulukhaktokmiut excel. This view of literacy, as monolingual fluency in the script of a national language such as English or French, dominates public discourse about literacy and institutional structures for literacy programs and research in Canada.

It was within this context that two proposals from Ulukhaktok—to offer literacy workshops in drum dancing and singing, as well as printmaking—were initially rejected. Emily Kudlak, the community language coordinator, asked, "Why is drum dancing and singing, or printmaking not literacy?" (Balanoff, Chambers, & Kudlak, 2005). The authors, both from outside Ulukhaktok, together with Emily Kudlak and Alice Kaodloak, both Ulukahktokmiut, formed a research team who met with various community representatives to seek support for the project. We proposed to document elders' knowledge of the pattern-languaged literacies of Ulukahktokmiut, as well as their experiences learning those literacies (Balanoff, Chambers, Kaodloak, & Kudlak, in press). The community approved the project because it proposed (1) to accrue economic benefits to Ulukhaktok and (2) to forestall the loss of *pitquhiraluavut* (our ways). Although formal protocol agreements were not negotiated (Brown & Peers, 2006), community representatives expressed confidence that the community's interest would be protected by the ongoing participation of the local researchers.

Planning for Participation

Cornwall and Jewkes (1995) suggest that even though "working with local people is far from easy" (p. 1673) there are principles, which if adhered to, ensure a project is more participatory than if conventional social science research methods alone are used. However, the meaning and practice of participation, like other forms of indigenous knowledges, are "generated in the practices of locality" (Ingold & Kurttila, cited in Huntington, 2005, p. 30); in other words, outside researchers, even when experienced, neither can nor should predetermine what participatory means or how it is practiced in a local context. In such contexts, it may be the attitude of the researchers, more than a set of research methods, that shapes how participatory a project actually becomes (Bishop, 1997, 1998; Cornwall & Jewkes, 1995). With knowledge of participatory research methods and a participatory attitude in hand, the authors planned for the ULRP "to be as participatory as possible, under the circumstances."[3]

The authors of this chapter are both from outside Ulukhaktok: one a former northerner now teaching at a Canadian university, the other a thirty-five-year northern educator working with a territorial literacy organization. The local co-researchers (Weber-Pillwax, 2004) have expertise with *pitquhiraluavut* (our ways), fluency in Inuinnaqtun, prior experience working with outside researchers conducting studies in the community, as well as, *ihuma* or Ulukhaktokmiut social intelligence (Stern, 1999). These critical skills enabled Emily Kudlak and Alice Kaodloak to provide project leadership in the community and to manage and carry out the research within the community. In most research projects previously conducted in Ulukhaktok the relationship between local "research assistants" and outside researchers (Fletcher, 2003, p. 52) has been primarily contractual. In the ULRP, however, Emily and Alice function as researchers, refer to themselves as researchers, and are referred to in the community, as researchers.

PAR projects vary in how equitably research benefits and potential for participation in decision making is distributed among participants, particularly between outside researchers and local people. Four possible modes of participation from least to greatest are contractual, consultative, collaborative, and collegiate (Cornwall & Jewkes, 1995). In the contractual mode, outside researchers control the agenda and hire locals primarily for specific, well-defined, nondiscretionary tasks. In the consultative mode, outside researchers seek local opinions, and perhaps political support, while retaining control and accruing most of the benefits of the research. In the collaborative mode, outside researchers work more closely with local people, while continuing to design, initiate, and manage the research. The ideal mode of participation, from a social justice standpoint, is collegiate where outside researchers and local people share a common purpose developed over an extended period of time. In a process of mutuality, they work together, each willing to learn from and listen to the other, and to contribute different skills to the common enterprise. Ideally, indigenous people "generate, analyse, represent, own and act on the information" sought in the research, thus theoretically making such projects "participatory as well as *participant-driven*" (Bishop, 1998, p. 204). This seems very much like Brigg's mode of collegiate participation.

The ULRP was designed to maximize meaningful participation and community ownership over the project. As Cornwall and Jewkes (1995) summarize,

> Ultimately, participatory action research is about respecting and understanding the people with and for whom researchers work. It is about developing a realization that local people are knowledgeable and that they, together with researchers, can work towards analyses, and solutions. It involves recognizing the rights of those whom research concerns, enabling people to set their own agendas for research and development and so giving them ownership over the process. (p. 1674)

However, how participatory a PAR project is rarely "follows the smooth pathway" implied in theory (Cornwall & Jewkes, 1995, p. 1672) or described in practice. As mentioned earlier, ideal modes of participation cannot be attained simply through careful construction of a participatory research design, or application of predetermined and prescriptive methodologies, not even with the saintly intentions of outside researchers (Rahnema, 1990). In the next section, we suggest this is so because participatory research occurs within a larger sociopolitical and institutional topography, one that does not provide the social conditions necessary for nurturing ideal modes of participation. In particular, the underlying assumptions, policies, and practices of the state, and its institutions for social science research, militate against socially just and locally appropriate modes of participation.

The Effects of Inadequate and Unfair Distribution of Resources on How Participatory a PAR Project Is

Social justice has been equated with the "morally proper" distribution of social benefits such as material goods and resources, and social burdens among a society's members (Young, 1990, pp. 15–16). Many contemporary philosophers of justice believe that recognition of entitlement to scarce resources and the proper allocation of those resources among a society's members are crucial to any understanding of social justice. In welfare capitalist states, like Canada, Young (1990) says, "public political dispute...is largely restricted to issues of taxation, and the allocation of public funds among competing social interests" (p. 19). Funding of research is part of this public debate in Canada: what constitutes legitimate social science research deserving of state funding? And, what constitute legitimate and reasonable research costs? If social science research—its outcomes and state funds to support those outcomes—is a "dominant good" (Walzer cited in Young, 1990, p. 17), then a monopoly on that good, by state institutions and the research paradigms or ideologies that dominate their decision making, is unjust. In the "distributive paradigm" (Young, 1990, p. 15), social justice demands a wider and more equitable distribution of the dominant good, whether that is money, land or resources for research.

In our experience, state policy and funding for community-based social science research in the North fails the fairness test. Even when research such as the ULRP receives state funding, the allocation of resources is not commensurate with the material needs and conditions of northern communities. Neither are those allocations appropriate relative to the resources required for projects to be participatory in ways that honor the practices of the communities where the research is situated. We offer four brief examples from the ULRP.

Building Research Capacity

First, institutions administering public social science research grants assume a hierarchy of merit and expertise. In this hierarchy, professional researchers mentor researchers-in-training and contribute to state goals to build research capacity through employment of undergraduate and graduate students as research assistants. No Ulukhaktokmiut are attending southern universities at present; few people in the community have a baccalaureate degree, and none have a graduate degree (to our knowledge). Therefore, there is no cadre of Ulukhaktokmiut available for mentorship through on-campus research assistantships. Besides, we believe that it is Ulukhaktokmiut ways of participating that must be expressed, utilized, socially recognized, and expanded through the social processes of the research. Therefore, it is training in being Ulukhaktokmiut that is essential to the research rather than training in Western social science research methods. Thus true research capacity building in northern communities means supporting Ulukhaktokmiut to participate in social science research in ways appropriate to Ulukhaktokmiut. As Young (1990) suggests, socially just participation means opportunities for groups, typically subordinated by asymmetrical relations of power, to develop and exercise their own capacities (for research) and to express their own experiences (and how those might be part of a research agenda). This means creating the conditions whereby community people can determine what actions constitute research and research methods in Ulukhaktok, and the conditions of those actions, in that context.

Basic Research Infrastructure

There is little or no research infrastructure within northern indigenous communities, such as Ulukhaktok. For example, the Quonset hut where the project is housed is only available through the goodwill of the local community government in cooperation with a regional cultural organization. Even then, the building is periodically without water and sewer and other amenities; and the project has few basic office supplies such as printing paper and ink.[4] Participation requires sustained communication and interaction among members of a research team, some of whom may live thousands of kilometers away. Although technology can facilitate virtual participation in between face-to-face meetings, most northern communities have neither the technology nor the broadband capacity to facilitate sustained communication through such technologies. Nor do they have the infrastructure to sustain maintenance and upgrading of such technology. This means that meaningful and sustained participation among research team members is restricted to conference calls (where local practices of participation may be audible but they are invisible) and air travel. Lack of basic research infrastructure in the North disadvantages participatory research. First, it makes it more difficult for northern PAR projects to compete for grants against professionals at universities with preexisting research infrastructure. Second, lack of funds for basic research infrastructure inhibits and constrains opportunities for a project to be maximally participatory. Social science research for northern indigenous communities needs facilities and technological infrastructure funding such as is available to natural scientists through research institutions such as NSERC (Natural Science and Engineering Research Council) in Canada and the National Science Foundation in the United States.

Face-to-Face Participation

State institutions allow expenditures for travel under tightly framed conditions of legitimacy. However, current funding parameters and envelopes (i.e., total maximums for specific programs) suggest blatant ignorance of basic facts about northern geography, as well as social and economic conditions: the distances to be travelled, and the time and costs involved. Airfare in the arctic is expensive, perhaps exorbitant; yet, it is the only efficient means of travel between many northern communities, and the only means of travel between arctic communities.[5] Given that professional researchers are required by the state to direct and manage research, a matter we will turn to later, and given that the ideal mode of being participatory in the North is direct, face-to-face meetings, insufficient funds for air travel and accommodation seriously inhibit conditions necessary for sustained participation in community-based social science research in the North.

Traversing the Linguistic Terrain

Fourth, state programs for social science research in Canada do not allocate sufficient resources for participatory research teams to traverse the arduous and complex linguistic terrain of cross-generational research in northern indigenous communities. Given state constructions of indigeneity, as difference within national homogeneity (Ingold, 2000), in Canada and in particular the NWT, indigenous languages are still considered a public good. A small percentage of limited state resources are expended to facilitate continued use and survival of the nine official indigenous languages in the territory. Federal research institutions allocate few resources for translation between indigenous languages and one of the official national languages; nor do they provide sufficient funds for conducting research in an indigenous language. Ideally, participatory research projects in northern indigenous communities ought to be conducted in the language(s) of the community at every stage of the research: to frame the research question; initiate and set the research agenda; and produce, construct, represent, and mobilize the knowledge. However, the language of empire, English—both in its standard dialectal variation (propagated by institutions such as schools) and in its localized dialectical variation (everyday colloquial speech)—continues to occlude the indigenous languages. And communities like Ulukhaktok may have speakers of more than one indigenous language. These linguistic complexities call into question a single, authoritative, and authentic standard of speech. Alphabetic systems for writing indigenous languages introduced by missionaries complicate the oral and aural variations of language use. Sometimes old and new orthographies for indigenous languages compete, but ironically a paucity of print materials, in either system, makes standardization of writing somewhat moot but still problematic for all stages of research.

Conducting participatory research in northern communities requires continual translation among languages, dialects, modalities, and discourses. Elders speak in the old language, which middle-aged researchers exert enormous effort to translate into English vernacular; outside researchers communicate with local researchers (both of whom come from differing interpretive communities and discourses within the contact zone) to interpret the translations; and the team as a whole struggles with the questions of what language (or dialect), discourse, and modality is the best in which to communicate with each other, the community, and the outside world, about the research. Researchers translate translations of translations, and interpret interpretations of interpretations (Nagy, 2006). Then they must translate these interpretations into discourse comprehensible to funders (Ryan, 1994), and other constituencies who call for their allegiances. Thus in the North, research, researchers, communities, and results are easily lost in translation (Hoffman, 1990). For indigenous community-based research projects, such as the ULRP, to aspire to ideal modes of participation, nation states need to recognize indigenous languages as part of the dominant good (for the entire nation state not simply for the speakers), and as worthy of receiving just allocations of state resources to make possible use of those languages for public activities such as research.

In summary, within the northern, and (post)colonial context, the possibilities for unfair distribution of resources is always present. Inadequate allocation of resources to research in northern indigenous communities constrains participation, and thus compromises the capacity of PAR projects to be socially just either in the social processes of the research or in its outcomes. In the next section, we extend the discussion beyond the distributive paradigm to how public policy and institutional discourses enframe social science research as professional work, thus inhibiting even more extensively the full and meaningful participation of indigenous groups in research done in their communities. We turn our attention to potential for abuses of power within the social processes of research, particularly when professionals and their institutions wield power unavailable to individuals and communities in the Canadian North.

Beyond the Distributive Paradigm: Power and Participation

Many contemporary liberal social theorists equate justice with the morally proper distribution of social benefits such as material goods and resources. However, efforts to redistribute goods that ignore the structural relations, which give rise to the underlying inequities, will not alone create social justice. While fairness in distribution of wealth is important for meeting the material needs of a society, extending the distributive paradigm from goods to social power is problematic (Young, 1990). Treating power like stuff assumes that it can be distributed in greater or lesser amounts much like money. Also the distributive paradigm places power with individuals, so individuals redistribute goods to individuals. These ideas however obscure the fact that "power is a relation" that exists "only in action" (Foucault cited in Young, 1990, p. 32). Power is not a thing that can be possessed by individuals and redistributed like goods. Young (1990) argues that individual agents—such as police, teachers, or researchers—only have institutionalized power over another if their actions are supported by a larger network of practices of power, such as a judicial or school system, or a national government research program. Therefore, in contemporary Western societies "widely dispersed persons are agents of power without 'having' it, or even being privileged" (p. 33) within those societies. Thus, social power is more than possessing and consuming. Therefore, there is social justice when persons have the opportunities to be "doers and actors" (Young, 1990, p. 37), in social arenas that are meaningful to them. Framed this way, socially just participation is

> learning and using satisfying and expansive skills in socially recognized settings; participating in forming and running institutions, and receiving recognition for such participation; playing and communicating with others, and expressing our experiences, feelings, and perspective on social life in contexts where others can listen. (p. 37)

While social justice is the realization of these values, social conditions, and institutional arrangements that constrain these capacities for participation are unjust. For Young (1990), two social conditions define injustice: oppression and domination. Oppression is the "institutional constraint on self-development," and domination is the "institutional constraint on self-determination" (p. 37). In this view, social justice is the realization of opportunities to participate meaningfully in social life; and social injustice is institutional constraint on those opportunities.

Who Has the Brains?

If inadequate and improper allocation of resources constrains the capacity of a project to be participatory, research institutions also constrain the capacity of indigenous groups to self-develop and to self-determine in the context of social science research. The policies and programs of granting agencies, such as the Social Sciences and Humanities Research Council (SSHRC) in Canada, insist that professional academics provide intellectual direction to social science research. Grant applications, which involve collaboration with nonacademic constituents, require evidence that a professional intellectual will direct the research and evidence of their competency to do so. Even in those SSHRC grants designated for Aboriginal research, the assumption remains that professional researchers will direct the project intellectually, although preference may be given to indigenous professionals. The reverse underlying assumption is noteworthy: intellectual direction for social science research is beyond the capacity of nonprofessional, nonacademic indigenous constituents. This is highly problematic for projects such as the one in Ulukhaktok where the very quest is for the intellectual material—that is *pitquhiraluavut* (our ways)—whose competency is called into question by these arrangements.

Who Has the Money?

Along the same nexus is financial accountability. State institutions that fund research dictate salaries, timelines, and reporting procedures. Moreover, these institutions require professional researchers, and their affiliated institutions, to hold and administer the grants, and thus by default "manage" the research agenda. These structural arrangements are in stark contrast to community ownership of research or indigenous peoples' understanding of participatory that assumes that

> attempts at developing symmetrical dialogue move beyond efforts to gather "data" and move towards mutual, symmetrical, dialogic construction of meaning, within appropriate culturally constituted contexts, [where] the voice of research participants is heard, and their agency is facilitated. (Bishop, 1998, p. 208)

Community-based organizations, without going through onerous and lengthy procedures to be approved by the state (in Canada, the Treasury Board) to administer grants, must rely on an institutional partner to apply for and manage the grant. The assumption underlying these structural arrangements is that nonprofessional constituents, particularly northern indigenous communities and organizations, are incapable of being fiscally responsible. Ironically, these demands for professional control over research are present even when specific grant programs require professional researchers to partner with community agencies and groups. This suggests state formulations of what partnership means when it is between professionals backed by public institutions (like universities) and nonprofessionals from indigenous communities and organizations. In these contexts, collaborative and partnership—highly potent terms in contemporary state discourses about social science research—really mean contractual or consultative rather than collegiate.

Who Is on Top?

Institutional policies on funding arrangements for community-based research both assume and reinforce underlying class relations and a hierarchy of labor. Professionals (and the middle class) are privileged in relation to nonprofessionals (and the working class). This hierarchy of labor constrains participation, particularly around decision making, in the taskscape (Ingold, 2000). As Young (1990) writes, "A hierarchical division of labor that separates task-defining from task-executing work enacts domination, and produces or reinforces at least three forms of oppression: exploitation, powerlessness, and cultural imperialism" (p. 12).

Exploitation is the unequal distribution of resources, which "is group based and structurally persistent" (Ackerman cited in Young, 1990, p. 52). It involves a steady "transfer of the results of the labor of one social group to benefit another" (Young, 1990, p. 49). Research-funding institutions allocate more resources and decision-making power to the professional class than the nonprofessional class, who in northern communities also happen to be indigenous. As long as national publicly funded research institutions and programs insist the research dollars (financial benefits) and the intellectual direction (power and control) remain in the tight grip of academics and their institutions, professional researchers affiliated with universities will continue to expand their careers and expertise, to accrue autonomy and respectability, and to make more money than contracted local nonprofessionals. Social justice requires "the reorganization of institutions and the practices of decision-making" (p. 53); in other words, social justice requires a different meaning and practice of participatory, one that is neither hierarchical nor class or race-based. This means reexamining what is at stake when funders of social science research assume a social division of labor between "those who plan and those who execute" (p. 58).

Whose Story and How Will It Be Told?

In the case of northern PAR projects, such as the ULRP, exploitation and powerlessness are potentially exacerbated by colonial constructions of race, where the community is almost always indigenous and the outside researchers typically are not. Although not specifically racist in their discourse, institutional funding arrangements for social science research, described earlier, are marked by what Young (1990) calls cultural imperialism. In this form of asymmetrical relations of power, the dominant group universalizes its own knowledge practices, and establishes them as the norm, while marking groups subordinated by these arrangements as Other, and their practices and cultural products as remarkable, different and, particularly in educational contexts, deficient. Through these arrangements indigenous peoples have been objects of study rather than agents of research, subjects of study to which knowledge is applied (by others), rather than agents actively generating knowledge themselves. Thus, discourse and practices generated from outside indigenous communities enframe what constitutes problems for those communities, for example literacy, success, and fiscal responsibility, as well as what solutions for those same problems might be and how they might look. The language, voice, and media of the products of research must be recognizable to the funders as outcomes or deliverables. This militates against indigenous discursive practices and modes of representation as constituting legitimate and sufficient products of research. One consequence of this intellectual imperialism is that while the products of social science research are normalized as knowledge, the products of research with indigenous communities are differentiated as perhaps downgraded to culture.

Conclusions

Indigenous communities in the Canadian North face numerous barriers to exercising their capacity to participate fully in all stages and aspects of research in their communities: lack of basic research infrastructure, prohibitive operating and travel costs, and complex and difficult linguistic terrain, all of which complicate and constrain participation, particularly between local and outside researchers. The state fortifies these barriers with federal requirements that professionals, academics affiliated with research institutions approved to administer state research grants, provide the intellectual direction for, and fiscal management of, the research, as well as assume responsibility for products and outcomes. When outside nonaboriginal researchers work with northern indigenous communities, these state conditions enframe the activities of participation of the research team. They assume and reinforce a hierarchical division of labor that separates the team into those that define tasks and those that execute them, which in the North also often divides power along lines of race and language. These institutional assumptions, and the funding arrangements they legitimate, constrain the capacity of indigenous communities to determine what participation means for research in their community and how it shall be practised; and they constrain the capacity of indigenous communities to develop their own research agenda, as well as their own legitimate and appropriate research methods based in *pitquhiraluavut* (our ways) or in "practices of locality" (Ingold & Kurttila, cited in Huntington, 2005, p. 30).

Social science needs to be socially just in its processes as well as its outcomes. In the Canadian North, this means that state institutions for funding social science research must provide the necessary resources as well as the social conditions and the institutional arrangements necessary to ensure meaningful and maximal participation of indigenous communities in research, particularly in research in their communities and spheres of nurture (Ingold, 2000). An important first step will be to recognize that activities of participation in indigenous communities are not exotic cultural practices or explicit sets of protocols amenable to translation and export. Neither are they deficiencies demanding remediation or erasure. Rather how a people participate in the world where they dwell with others is knowledge and their activities of participation are the knowledge practices generated within, appropriate to, and necessary for, that locale. A socially just state program for research would provide the necessary social conditions for locally appropriate forms and ways of being participatory; it would support the practice of *pitquhiraluavut* (our ways) as knowledge preservation, generation, and creation. A necessary second step would be to make possible the exchange of this knowledge and these practices among northern indigenous communities and beyond. In this way, the benefits of knowing how to participate with each other in particular places will hopefully translate beyond research and past the circumstances of the local, to all northerners and southerners and the places they call home.

Notes

1. *Ulukhaktok* means "there are many good rocks to make *ulu*." (Collignon, 2006, p. 9); an *ulu* is a woman's knife with a semicircular blade. Originally named Holman, the community was officially renamed *Ulukhaktok* on April 1, 2006.
2. Most of the Inuinnait (a collective noun for twenty-one Inuit groups indigenous to the Central Arctic) who now live in Ulukhaktok are either *Kangiryuarmiut* ("people of the big bay," i.e., Prince Albert Sound) or *Kangiryuaqtiarmiut* ("people of the smaller bay," i.e., Minto Inlet) (see Collignon, 2006).
3. This is a play on "as Canadian as possible under the circumstances," one definition of what it means to be Canadian (Chambers, 2003).
4. Supplies are purchased locally at inflated costs (e.g., a ream of paper costs $19), or shipped in as air freight, also very expensive.
5. Return airfare from Yellowknife NWT to Ulukhaktok is $2200 CND.

References

Absolon, K., & Willett, C. (2004). Aboriginal research: Berry picking and hunting in the 21st century. *First Peoples Child & Family Review: A Journal on Innovation and Best Practices in Aboriginal Children Welfare Administration, Research, Policy and Practice, 1*(1), 5–17.

Balanoff, H., Chambers, C. M., & Kudlak, E. (2005). Do my literacies count as literacy? An inquiry into Inuinnaqtun literacies in the Canadian North. *Literacies, Research in Practice, Practising Research, 6*, 18–20.

Balanoff, H., Chambers, C. M., Kaodloak, A., & Kudlak, E. (in press). "This is the way we were told": Multiple literacies in Ulukhaktok, Northwest Territories. In M. Therrien & B. Collignon (Eds.), *Proceedings of the 15th Annual Inuit Studies Conference*, Paris, France.

Bishop, R. (1997). Maori people's concerns about research into their lives. *History of Education Review, 26*, 25–41.

———. (1998). Freeing ourselves from neo-colonial domination in research: A Maori approach to creating knowledge. *Qualitative Studies in Education, 11*(2), 199–219.

Brown, A. K., & Peers, L. L., with members of the Kainai Nation (2006). *Pictures bring us messages Sinaakssiiksi Aohtsimaahpihkookiyaawa: Photographs and histories from the Kainai Nation.* Toronto: University of Toronto Press.

Chambers, C. (2003). As Canadian as possible under the circumstances: A view of contemporary curriculum discourses in Canada. In W. Pinar (Ed.), *Handbook of international curriculum research* (pp. 221–252). Mahwah, NJ: Lawrence Erlbaum.

Chambers, C., & Hasebe-Ludt E., with Leggo, C., Oberg, A., Donald, D., & Hurren, W. (2007). Métissage. In J. G. Knowles, & A. Cole (Eds.), *Handbook of the arts in qualitative research: Perspectives, methodologies, examples and issues* (pp. 141–154). Thousand Oaks, CA: Sage.

Collignon, B. (2006). *Knowing places: The Inuinnait, landscapes and the environment* (L. W. Muller-Wille, Trans.): Canadian Circumpolar Institute. Edmonton, Canada.

Cornwall, A., & Jewkes, R. (1995). What is participatory research? *Social Science and Medicine, 41*(12), 1667–1676.

Fletcher, C. (2003). Community-based participatory research relationships with Aboriginal communities in Canada: An overview of context and process. *Pimatisiwin: A Journal of Aboriginal and Indigenous Community Health, 1*(1), 28–62.

Heshusius, L. (1994). Freeing ourselves from objectivity: Managing subjectivity or turning toward a participatory mode of consciousness? *Educational Researcher, 23*(3), 15–22.

Hoffman, F. (1990). *Lost in translation.* New York: Penguin.

Huntington, H. P. (2005). "We dance around in a ring and suppose": Academic engagement with traditional knowledge. *Arctic Anthropology, 42*(1), 29–32.

Kritsch, I. (2002). The Gwich'in traditional caribou skin clothing project: Repatriating traditional knowledge and skills. *Arctic, 55*(2), 205–210.

Legat, A. (1995). Participatory action research in Rae Lakes, NWT: The traditional government project. In *NWT Diamonds Project environmental impact statement: Volume I—Appendices / BHP Diamonds Inc. and DIA MET Minerals Ltd.* Vancouver, BC: BHP.

Lionnet, F. (1989). *Autobiographical voices: Race, gender and self-portraiture.* Ithaca, NY: Cornell University Press.

Nagy, M. (2006). Time, space and memory. In P. Stern, & L. Stevenson (Eds.), *Critical Inuit studies: An anthology of contemporary arctic ethnography.* Lincoln: University of Nebraska.

Pratt, M. L. (1992). *Imperial eyes: Traveling writing and transculturation.* New York: Routledge.

Rabesca, M., Romie, D., Blackduck, R., Zoe, S., Legat, A., Johnson, M., & Ryan, J. (1993). Participatory action research workshop. Presented at *Human Dimensions of Northern Research*, Arctic College, Fort Smith, October 2. Retrieved from http://136.159.147.171/scripts/minisa.dll/627/2/11?RECLIST

Rahnema, M. (1990). Participatory action research: The "last temptations of saint" development. *Alternatives, XV*, 199–226.

Rasmussen, D. (2000). Our life out of balance: The rise of literacy and the demise of pattern language. *ENCOUNTER: Education for Meaning and Social Justice, 13*(2), 13–21.

Ryan, J. (1994). *Traditional Dene medicine: Part one report.* Calgary, AB, Hay River, NT, Lac La Martre: Arctic Institute of North America, Dene Cultural Institute, Lac La Martre Band.

———. (1995). *Doing things the right way: Traditional justice in Lac La Martre, N.W.T.* Calgary, AB: Arctic Institute of North America.

Sinclair, R. (March, 2003). PAR and Aboriginal epistemology: A really good fit. Retrieved from *Aboriginal & Indigenous Social Work*, Special Topics http://www.aboriginalsocialwork.ca/special_topics/par/epistemology.htm

Sissons, J. (2006). *First peoples, first cultures and their futures*. London, UK: Reaktion Books.

Stern, P. (1999). Learning to be smart: An exploration of the culture of intelligence in a Canadian Inuit community. *American Anthropologist, 101*(3), 502–514.

Thorpe, N., Hakongak, N., Eyegetok, S., & the Kitikmeot Elders. (2001). *Thunder on the tundra: Inuit Qaujimajatuqangit of the Bathurst caribou*. Vancouver, BC: Tuktu and Nogak Project.

Watkins, M. (Ed.). (1977). *Dene nation, colony within*. Toronto: University of Toronto Press.

Weber-Pillwax, C. (2004). Indigenous researchers and indigenous research methods: Cultural influences or cultural determinants of research methods. *Pimatisiwin: A Journal of Aboriginal and Indigenous Community Health, 2*(1), 77–90.

Young, I. M. (1990). *Justice and the politics of difference*. Princeton: Princeton University Press.

CHAPTER SEVEN

Action Research for Curriculum Internationalization: Education versus Commercialization

ROBIN MCTAGGART & GINA CURRÓ

Participatory Action Research as Social Practice

Participatory action research (PAR), more than a research methodology (Carr, 2006), brings people together to reflect and act on their own social practices to make them more coherent, just, rational, informed, satisfying, and sustainable. PAR involves groups of people working together on a thematic concern (Kemmis & McTaggart, 1988) that arises in their practice. Kemmis and McTaggart (2005) described public spheres, where people gather to develop shared understandings about the issues they confront and how to address them. It is not usually research per se that engages people, but working together, changing their practice in informed and responsible ways, developing concepts for discussing their work and collaborating differently with colleagues.

PAR involves studying, articulating, and extending another social practice simultaneously. Analysis of the subpractices of PAR illustrates its integration with another consonant activity worthy of the name practice (Schatzki, 1996). In this sense, PAR is a practice-changing practice (Kemmis 2007, October). The subpractices include

- research conducted by participants around shared concerns arising in their practice, research designed to inform action by participants, conceived differently from conventional research but sharing some aspirations—the objectification of experience (leading to the extension of the critical theorems in Habermas's [1974] terms);
- self-reflection by participants about their concerns arising from the practice—disciplining subjectivity in the affective sense (leading to the organization of enlightenment);

- self-reflection by participants about what to do in their respective situations—disciplining subjectivity in the political or agential sense; (the conduct of the political struggle); and
- collaboration among participants that makes these activities possible, disciplining the community of practice engaged in PAR, and changing practice (formation of public spheres in Habermas's [1996] terms).

The commercial impulse to recruit international students threatens to outrun educational aspirations for cross-cultural dialogue, critique, and collaboration. We aim to provide specific concepts to contest antieducational values expressed by the neoliberal ideology of transnationalism (Lindblad & Popkewitz, 2004; Rapley, 2004; Rizvi, 2004). We argue that PAR among professors[1] for curriculum internationalization means changing: (1) communicative practices (language, symbols, and other representations of teaching), (2) organizational practices (relationships among teachers and learners), and (3) work practices (teaching, management, teacher education, and evaluation) (Kemmis & McTaggart, 2000). Educational change emerges in public spheres (Kemmis & McTaggart, 2005) created by participants to generate communicative space. We describe curriculum internationalization at technical, practical, and critical levels (Carr & Kemmis, 1986; Habermas, 1974, 1987) to provide a heuristic for people changing their practices.

Public spheres provide forums for disciplining changes in educational work, but universities must create the conditions for informed and collective reflective practice. Internationalization must explore how (1) professors and support professionals learn from each other; (2) curriculum, teaching, and learning in classrooms develop; (3) organizational practices evolve; and (4) research, evaluation, and theorizing practices inform and influence teaching and learning. Participants need to consider the interactions between these practices and individual subjectivity (skills, values, understandings), social structures (culture, economy, political life), and social media (language, work, power). Changes fundamental to curriculum and PAR are expressed in all these domains.

A Thematic Concern: Internationalization of Curriculum

A thematic concern describes inchoate dissatisfactions that bring people together. Dissatisfactions are articulated, understood, and felt more deeply through dialogue. Internationalization exemplifies educational concern, becoming an institutional commitment long before professors comprehend its impact. It troubles professors whenever university management pleads commercial necessity before considering educational commitments. Issues include the following:

(1) Ethnocentric interpretations of the capacities of students whose first language is not (standard) English: Some professors believe

that Confucian Heritage Culture (CHC) students are rote learners. However, CHC students perform better than Australian peers on assessment tasks that require significant mastery of deep learning. CHC students cooperate in spontaneous learning groups in preparation for assessment tasks. Compared with competitive and individualistic styles where students do not share feedback, cooperative learning enhances student achievement (Boud, Cohen, & Sampson, 2001; Ho & Chiu, 1994; Ramsden, 2003; Watkins & Biggs, 1996). CHC learning approaches lead to deep learning, not just rote learning (Kember, 2000; Watkins & Biggs, 2001). Still, academic performance of CHC learners is ignored in favour of simplistic generalizations about their propensity for rote learning.

(2) Plagiarism allegations directed at students whose first language is not English: This is immensely controversial with blame too easily attributed to students. Entry levels for English competency are set low, rarely meeting language and writing requirements of courses. Competence for undergraduate entry is usually considered to be about IELTS level 6.5. Undergraduate proficiency generally requires an IELTS around nine. IELTS authors estimate that a gain of 0.5 in level requires one hundred hours of instruction from specialist teachers. Rarely does this happen.

Challenges facing international students for whom English is a second language are well documented, but support programmes for students remain patchy and poorly integrated in curriculum. Formal qualifications and teaching expertise to develop language skills in the students' disciplines are thinly spread. Inadequate practice persists despite documented success of integrated genre approaches in second language learning (Cargill, Cadman, & McGowan, 2001; Swales 1990, 2004; Swales & Feak, 1994). Accordingly, the required rates of improvement are unrealistic. Language support for students also lags behind punitive practices to curb plagiarism (Lindsay, 2008). Predictable concerns about English fluency of graduates have emerged (Birrell, 2006; Bretag, 2007).

(3) Impact of neoliberalism and economic rationalist culture (Pusey, 1991), the performativity problem in working lives of professors: Australian education is beset by a discourse of outcomes[2] and their measurement as indices of quality, intensifying the performativity culture (Altbach, 1991, Keenoy, 2004; Marshall, 1999; Vidovich & Porter, 1999). Professors perceive students from different language backgrounds as yet another demand in increasingly monitored lives. Professors are not just feeling overloaded, but hostage to "the intensification of world-wide social relations" (Giddens, 1990, p. 15). Discomfiture is compounded by contradiction: globalization impedes education for globalization.

Professors are also told that simple measures of quality are necessary for competitive markets to function properly (Brown, 2007). This ideology militates against reflective work on curriculum internationalization, despite international, national, and university policy and advocacy (Aulakh, Brady, Dunwoodie, Perry, Roff, & Steward, 1997; Butorac, 1997; Haarlov, 1997; Hooper, Garnett, Walsh, Vicziany, Rizvi, & Webb, 1999; IDP Education Australia, 1995; OECD, 1994; Rizvi, 2000). Still, there may be communicative space for people dissatisfied with progress on authentic internationalization: Leask (2001) reports glowingly about her university's progress. Activist professionalism (Avis, 2005) may yet express postperformativity.

Whatever the motives of advocates, internationalization is destined to concern professors. To support changes in practice among professors and other university professionals, we present a heuristic derived from the literatures of internationalization, curriculum, and PAR. The heuristic recognizes that institutional context and daily interactions between learners, teachers, and curriculum resources matter. It follows that the key change agents are professors as the primary community of practice (Wenger, 1998). Professors are not implementers of internationalization, but creators of it over time, with support from policy, other professionals, and students.

Internationalization implies curriculum change in two ways:

(1) making the curriculum more engaging and relevant for students from cultures different from that of the university and
(2) preparing students from many cultures to live and work in settings different from the university's home culture.

Internationalization affects all fields of university study in some way. Many fields are affected explicitly by an influx of international students, on-campus, online, and offshore. Most universities enrol students from increasingly diverse cultural backgrounds. Accordingly, students have different expectations about relationships between teacher and learner, knowledge and authority, critique and disrespect, and buyer and seller. Universities also have domestic students needing help to develop the skills, understandings, and values necessary to work in organizations, systems, and communities different from those they know.

Cross-cultural issues apply even where non-Western students attend Australian universities to learn the dominant scientific and technological discourses that define Western culture. It cannot be argued that no change is necessary because science or technology is universal. All knowledge is reinterpreted in particular contexts, and workplaces; customs and industrial laws fall well short of global uniformity. The location, shape, purpose, and scale of a bridge are aesthetic, social, and cultural matters, not merely engineering problems. All disciplines derive from diverse global traditions and any disciplinary resource for university education should recognize that historical, cultural, and material formation. No discipline is exempt from the need to reflect and change. Social technologies such as enrolment procedures too should heed expectations of students from different cultures.

Obviously professors are changing aspects of their teaching most of the time, but this would not yet be a form of PAR. Not all changes in practice are informed by relevant theory, exemplary cases, and collective inquiry in the situation. Justifiable changes in practice involve working with colleagues, responding to the disciplines of data, knowledge of teaching, and prudence in deciding how to improve. Orlando Fals-Borda (1979) called participatory research investigating reality in order to transform it. Later, he affirmed the importance of the complementary phrase transforming reality in order to investigate it (Fals-Borda, 1989). Freire (1982) argued that the only way of creating alternative research methods, meaning PAR, was learning to do it by doing it.

Changing a practice becomes PAR when approaches consistent with the literatures of the field, innovation, change, and critical social science are used. More justifiable practices imply theoretically informed ways of transforming them. Internationalization takes time and can occur at different levels of engagement. It begins with matter-of-fact technical changes, often educationally unsatisfying. Communities of practice expand among professors acting on their dissatisfactions, embarking on a continuum from technical observance, through practical deliberation and deeper participation, toward critically reflective practice.

Direction for Internationalization—From Technical Observance to Critical Reflection

Technical observance of a commitment to internationalization is illustrated by an emphasis on

(1) recruiting more international students,
(2) employing staff with international backgrounds and experience,
(3) using international examples in curricula,
(4) adding support services to help students survive, and
(5) seeing "poor" English as a clinical condition requiring remedy.

In the technical view, internationalization is an add on. Students are meant to adapt to the university expectations, inviting cultural imperialism and viewing students from a deficit, ethnocentric perspective (Biggs, 2003; Volet, 1999). Education involves submitting to the authority of legitimated texts; there is cultural reproduction in all education. However, university education also involves cultural production, requiring a dialectical relationship between text and learner, teacher and taught, milieu and student. International education means recreation of globalization as social practices to build humane transcultural lives by renegotiating the practices of

(1) curriculum, teaching and learning, and learning support,
(2) organization and administration as the context for social relations,
(3) staff development (academic and general), and
(4) research into, and evaluation of, curriculum, teaching and learning, and learning support.

The goal of this renegotiation has three elements (Rizvi, 2000). First, it implies a curriculum that equips students to critique and engage the knowledge economy; to understand local and global influences on knowledge production, distribution, and use. Second, curriculum must draw on the global plurality of knowledge sources, links, and influences to encourage local use and global participation. Student understanding must be extended by global imagination. Third, universities must illustrate their distinctiveness as social institutions by valuing openness, tolerance, and cosmopolitanism, creating an inventive learning ethos, engaging cultural differences, and developing perspectives that allow comparisons and critique. In short, internationalization involves reviewing and changing most things a university does.

Change may begin with some practices at the technical level, partly because universities have much to learn about educating internationally. A serious educational problem arises if practices remain at a technical level. From the critical perspective, technical procedures are only acceptable as developmental steps toward critically reflective practices. Vigilance is necessary to ensure that development moves the university toward an authentically critical internationalist view of itself. Liberal ideas like cultural sensitivity may despatch racism, but still denote otherness whereas cultural inclusiveness reflects confidence, reciprocity, and critique to create inclusive curriculum. It means including people, and also including different cultural perspectives in the curriculum, attempting to construct dialogue between cultures that does not homogenize, denigrate or romanticize them.

How do professors decide their preferences in international inclusive, critical curriculum practice? What is the content of their conversations? At the individual level, changing teaching and curriculum involves three basic areas. Professors need new skills. They also need new understandings to justify different curriculum content and teaching practices, and revised values to articulate and guide their work. These extend existing repertoires and emerge through interaction with university and discipline colleagues. How do we inform and examine those changes at the individual level? How do we help people to see their practices in different ways?

Concepts from PAR

Kemmis and McTaggart (2000, 2005) provide a framework useful for the internationalization of pedagogy. Practice can be viewed in several ways:

(1) the individual performances, events, and effects that constitute practice as it is viewed from the "objective," external perspective of an outsider (how the practitioner's individual behaviour appears to an outside observer);
(2) the wider social and material conditions and interactions that constitute practice as it is viewed from the objective, external perspective of an outsider (how the patterns of social interaction among those involved in the practice appear to an outside observer);
(3) the intentions, meanings, and values that constitute practice as it is viewed from the "subjective," internal perspective of individual practitioners themselves (the way individual practitioners' intentional actions appear to them as individual cognitive subjects);
(4) the language, discourses, and traditions that constitute practice as it is viewed from the subjective, internal social perspective of members of the participants' own discourse community who must represent (describe, interpret, evaluate) practices in order to talk about and develop them, as happens, for example, in the discourse communities of professions (how the language of practice appears to communities of practitioners as they represent their practices to themselves and others); and
(5) the change and evolution of practice, considering all four of the aspects of practice just mentioned, that comes into view when it is understood as reflexively restructured and transformed over time, in its historical dimension.

The first four of these lead to familiar research approaches and techniques. Our interest is the fifth perspective that creates challenges by being more than a research approach—reconstitution of practice through informed human agency. The goal is the immediate and continuing betterment of practice rather than merely being informed about practice. Since changing practice is the focus, we must put ourselves into the workplace and consider what kinds of information we and others might need. We need to take into account not just what people might think about the current situation, but how they might respond if we begin to initiate changes. This requires an understanding of individual views and shared social understandings. Even objectively established facts such as the number of students who speak languages other than English in a class will involve different subjective reactions. The individual, social, objective, and subjective perspectives in the situation must be taken into account, if we are to do something.

In one sense, the fifth perspective takes an aerial view of the four other approaches, and instead of fragmenting into each of the four respective specializations of method, it considers them together. It engages the kinds of questions each perspective addresses, but in a somewhat different way. It does not anticipate the distillation of a study of the situation, but instead concentrates on changing participants' understandings, their practices, and the situation in which these are constituted. Each of these, understanding, practice, and situation have been formed in particular historical, material, and political settings, and it is theoretical insight from critical social science that guides reflection and action. This was a view of PAR proposed by Kemmis and McTaggart through the 1980s (Kemmis & McTaggart, 1988). Fortuitously rather than presciently, it reflects shifting views of the relationship between philosophy and life and theory and practice emerging in the late twentieth century (Hadot, 1995; Kemmis, 2007; Schatzki, 1996, 2002).

Consider the ways in which social practices are constructed and contextualized. We conceptualize this at the *individual* level in terms of knowledge and social practices, and at the *social* level in terms of social structures and social media. As individuals embark on a change in practice (beginning at the top of table 7.1), their state of *knowledge* will change as they develop new understandings from their reading and dialogue with others. They will acquire new skills, and it is likely also that their values will change as they learn. Changes in their own *social practices* will be constituted through new ways of communication within the practice (with their students) and about the practice (with students, colleagues, and others). Their practices of production (teaching, curriculum development, assessment) will change, as will patterns of organization (relationships with students, parents, and others). What is achieved will be a function of the *social structures* in place. The culture of the university, faculty, classroom, and staffroom will provide both opportunities and constraint for change, but will also be amenable to change itself, perhaps necessarily if changes are to be effected and embedded. The economy of the setting will respond similarly—how resources are distributed and how people spend their time and emotional energy will be key influences on what can be accomplished. Changes in political economy of information production and distribution will exert an impact here too. In turn, these will engage current and possible forms of political life. Participants come to these structures as sedimented practices, but they are not fixed and can be changed to effect educational change. The *social media* of language, economy, and political life (sayings, doings, relatings) are the ways in which changes in professional practice find expression. New ways of thinking and saying things about practice will be associated with changed ways of working. These in turn are linked with emerging forms of relationship that signify the possibility of more satisfying and sustainable forms of educational life.

Table 7.1 Change in the domains of practice

	Knowledge		
Individual (subjectivity)	Understandings	Skills	Values
	Social Practices		
	Communication	Production	Social Organization
	Social Structures		
Social (structure, ideology)	Culture	Economy	Political life
	Social Media		
	Language (discourses)	Work	Power

Because we are focusing on pedagogy for communities of practice, our initial interest is the domain of *social practice*. Within the domain of social practice, we have suggested that pedagogical innovation requires change in three aspects of practice:

(1) *communication* (language, symbols, other representations of educational work),
(2) *production* (teaching, leadership and management, teacher education, research and evaluation), and
(3) *organization* (relationships among teachers, learners, managers, and others).

Further, we suggest that educational practice consists of four subpractices: curriculum for students; administration; curriculum for professor development; and educational research and evaluation. Please refer to table 7.2 for a description of these subpractices.

Practices and subpractices embody common ideological roots and implicit understandings, so they are related and overlap. The mindset of practitioners is constituted in similar ways. There are shared understandings at the conscious level, shared assumptions in the subconscious. Note that the subpractices are not simply the practices conducted respectively by professors, administrators, staff developers, and researcher/evaluators. Elements of all of these subpractices are evident in all of these forms of work. For example, teachers do curriculum work, but are also involved in

- formulating policies for their own classes and the university generally,
- administration and organization,
- teaching colleagues as well as students, and
- researching, theorizing, and evaluating their own educational work.

Understanding relationships among practices is as important as understanding the relevance of relationships among people engaged in different

Table 7.2 The Sub-practices of university education

Media Subpractices	Communication	Production	Organization
Curriculum	Curriculum content of university courses	Pedagogy	University classroom authority and control
Administration	Educational policy	Administrative practices	Relations of authority and evaluation—university with staff
Teacher Development Curriculum	Curriculum content of staff development programs	Pedagogy	Staff development classroom authority and evaluation
Educational Research & Evaluation	Educational theory	Internal research and organizational evaluation practices	Politics of internal research and organizational evaluation

work. For example, assessment practices of professors are influenced by university administrative deadlines about the processing of assessment results. Both are preempted by system administrative (and legal) requirements like immigration departments monitoring student attendance and progress. Practices are intricated and assume common conceptual features.

The Heuristic Articulated

Table 7.3 includes in each cell a gradation of practice from the technical to the critical, to suggest the preferred developmental direction. Table 7.3 also

(1) provides concrete examples of different kinds of practice that will be recognizable to different participants. This should enable participants to plan things to try out in their own settings, discuss the rationale for the different kinds of practice and their adequacy, and begin a shared critique of concrete practices using some shared language and understanding;
(2) illustrates the ways in which the categories make certain practices visible (noting this can de-emphasize other practices such as students forming their own learning groups); and
(3) suggests how attention to relationships is necessary for the development from the technical to the critical to occur. Practices are manifold, not simple aggregations of separate activities. It is difficult to change one activity without effecting change in others. Changes must be studied and acted on in concert.

Use of the heuristic as practice changes over time allows PAR to flow. The discourse of PAR becomes more explicit as the work evolves. We have focused on a curriculum about practice as a first step for action researchers

Table 7.3 Examples of curriculum internationalization

University Educational Practices	Communication	Production	Social Organization
Curriculum	**Curriculum content**	**Pedagogy**	**Classroom authority and control**
Technical	Acknowledgment of international authors in the field. Language support available on student request—deficit approach	Correct pronunciation of student names	Polite Western manners
Practical	Teaching examples from other cultures. Language support available in one or two subjects—discipline study skills approach	Deliberate questioning of students from different cultures to sample perspectives	Consultation with students about preferred classroom relationships
Critical	Competing cultural explanations and perspectives presented for students to compare; language development integrated through whole curriculum—professor and language expert collaboration	Curriculum changes in dialogue with students to reflect cultural preferences and to nurture collaborative learning practices for deep learning and cross-cultural critique	Use of foreign language forms of address, idioms and patterns of respect and deference
Administration	**Educational policy**	**Administrative practices**	**Relations of authority and evaluation**
Technical	Policy emphasizes equality of opportunity	Service staff trained to deal with students from different cultures, especially students from non-English speaking backgrounds	Staff invited to undertake cultural sensitivity training
Practical	Policy emphasizes staff development for cultural sensitivity, inclusiveness	Staff of different cultures are appointed	Cultural sensitivity training required for all staff in performance management
Critical	Policy emphasizes representation of alternative cultural perspectives in all university practices	Active recruitment and support of academic and general staff to move ethnicity profile of staff toward that of students	Policies and practices systematically and regularly evaluated by stakeholders including community groups

Continued

Table 7.3 Continued

University Educational Practices	Communication	Production	Social Organization
Teacher Development Curriculum	Curriculum content	Pedagogy	Classroom authority and evaluation
Technical	Basic routines for involvement of all students in classroom interaction included	Participation in staff development for internationalization voluntary	"Oh that's interesting," "This is how I do it" classroom dialogue
Practical	Hierarchy of inclusiveness technical to critical/emancipatory included	Teachers explore and document internationalization practices extant in curricula they teach	Teacher educators encourage critique of teachers' ideas
Critical	Active critique and reform of curriculum required as part of course assessment	Teacher educators actively teach critique and assist teachers to change teaching and curriculum practices	Data about teaching is presented and improvements to teaching and curriculum planned, implemented, and evaluated critically
Educational Research and Evaluation	Educational theory	Research and evaluation practices	Politics of research and evaluation
Technical	Policy regarded as (1) binding aside from occasional review, or (2) irrelevant or too vague to guide practice	Evaluation of teaching dominated by use of standardised student rating scales	Research and evaluation practices are quantitative, *about* people rather than *for* them
Practical	Theories of teaching and learning well-documented but regarded as eclectic mix and largely a matter of individual personal choice and preference	Research on teaching is phenomenographic, qualitative with a view to discovering general principles of effective teaching	Research and evaluation tend to focus on the curious, the interesting and representation of the other
Critical	All university teaching and learning policies are treated as theories to be tested in practice and subjected to collective critique to improve practice by making it more rational, just, coherent, satisfying, and sustainable	Research and evaluation practices focus on teaching, curriculum, and the educational milieu together. These are disciplined by relevant literatures, collective critique, disciplined and informed self-reflection and commitment to improved educational practice	Research and evaluation seen as socially- and historically constituted practices and therefore correctable by stakeholder participation and a commitment to open and reasonable dialogue among those involved and affected

so that they recognize the depth and scale of the task they are embarking on. Without a mud map of the likely journey, people may get lost in the minutiae of technical action research.

Supporting Internationalization—An Extended Project

How can we provide students with a more inclusive and relevant education? How does the framework suggest a university should respond to stimulate and support internationalization of the curriculum? We know that professors interested in internationalization expect comprehensive rationales to justify changes in educational practice. Inconsistency among policies, practices, and leadership leads to rejection of innovation through lack of substance and commitment. Supporting change is therefore not easy. Educational institutions are loosely coupled (Weick, 1976); people are not readily directed to implement change (Fullan, 1989). Some are resistant to change, though among professors the contrary is typically the case. What nurtures change toward internationalization, and especially toward more critically reflective approaches to it?

The first suggestion is, start somewhere! The second is, persist! The examples within the *social practice* domain show different levels of critical engagement. Movement to critical change can also be achieved in each of the other domains, *individual*, *social structures*, and *social media*. It is easy to imagine the changes anticipated in the *individual* (subjectivity) domain. Individual professors should develop new understandings of the languages and cultures of their students to ensure that teaching and research supervision effectively align with the existing learning of students. Professors can improve their curriculum design, teaching, and assessment skills to help a more diverse student body, and recognize new dimensions of diversity. Some new skills are difficult to learn. Inviting an exchange of cultural perspectives and moving beyond presentation to comparison, critique, and reconciliation is a formidable and worthy task. Individual learning by professors also means changing values and their expression in practice.

As well as changing individual practices, we must work collectively in other ways to change *social structures* (culture, economy, political life) that create or constrain possibilities for curriculum internationalization. Internationalization involves changing institutional culture. A step in this direction means adopting internationalization as a project for the whole curriculum, not merely a curriculum for international students. This directs attention away from assimilation approaches based on ethnocentric, deficit models focusing on remediation (Volet, 1999). Institutional cultures embrace more than teaching and learning. Internationalization permeates institutional life: research and research training, staff recruitment, corporate support, clearly stated and shared aspirations, community relationships (service, input, advocacy), support for students, and critique, for example, in course and department reviews.

The material economy of institutional life must reflect commitment to internationalization. Political and material economy must influence educational expressions of internationalism ahead of its commercial utility in disciplines, faculties, and the university. People can be promoted for being good at curriculum internationalization, given time release to recast curricula, and teaching and staff development awards can feature exemplary curriculum reform. Bilingual staff should be recognized and staff well-qualified in applied linguistics and teaching English as a second language appointed and rewarded. In the current economic climate, failure to attract international students may lead to financial difficulties as some disciplines can generate resources for (and from) internationalization more readily than others. The university may develop budget approaches that share operating surpluses so that all disciplines can participate in curriculum internationalization. Commercial values should not dominate decisions about the curriculum profile of the university.

Closely related to economic influences is the real politik of institutional life. Not only material imperatives influence the way people act. Recognition, legitimate authority, and encouragement to innovate are important. The university may support staff wanting to internationalize against the conservative edicts of professional bodies (but recognize their capacity to stimulate innovation too). It makes a difference if the management of the university sponsors internationalization and works to create a culture of disapproval of monoculturalism. Most importantly it directs the political life of the university toward more satisfying and sustainable educational activities among its professors.

PAR creates new ways for university educators to talk about their work, to do their work, and to relate to each other about their work. The *social media* of language, work, and power come to express curriculum internationalization. However, practice architectures create several mediating preconditions for practice:

(1) cultural-discursive preconditions that shape and give content to the thinking and saying that orient and justify practices;
(2) material-economic preconditions that shape and give content to doing of the practice; and
(3) social-political preconditions that shape and give content to the relatings involved in the practice (Kemmis & Smith, 2008).

Understandings, practices, and the conditions of practice are bundled together (Schatzki, 2002) in manifold ways that are comprehensible to professionals, essential to working together, but tacitly understood and emotionally engaged. At one level practices are remade incessantly, but remaking practices at a deep level requires significant effort. Curriculum internationalization involves such effort, not least because institutional message systems compound the difficulty at deep levels through influencing:

- allowed and preferred discursive forms (e.g., favoring commercial over intellectual values);
- what counts as legitimate work (e.g., publication is more important than rewriting assessment tasks to increase their accessibility for non-English speaking background (NESB) learners); and
- patterns of deference that illustrate power differentials (e.g., a learning adviser with TESOL teaching skills deferring to a lecturer whose teaching is confusing international students).

Supporting the internationalization of pedagogy is a comprehensive task. It involves opportunities for academic staff to experiment with teaching and curriculum, and especially to respond to the learning practices already in students' repertoires. Internationalization can occur in simple technical ways or can be developed from a critical perspective with commitment to inclusiveness. None of this will happen without remaking social structures as well as practices themselves. Attention must be paid to the real politik and material economy of institutional life to make internationalization educationally satisfying for staff. Staff development must promote new ideas about educational practices, and professors must be given realistic opportunities to try out ideas sustainably. The university must sponsor disciplined and informed reflection on current practices and intended changes, which is the role of PAR. In this way, the university can internationalize its curriculum. When staff learn how to be participatory action researchers, they can do it better.

Internationalization is a watchword of change in Australian universities. It signals a commercial impulse, but at the same time expresses a commitment of Australian higher education to engage with the international community, reflecting a spirit of internationalism. Australian universities are becoming less homogeneously Australian and aspire to be global universities offering curricula and research programs with an Australian perspective. Just what this means is difficult to specify because internationalization will always be the subject of planned invention, experimentation, serendipity, reflection, and development over a period of time. How we move from advocacy and policy to create this innovative pedagogical work provides an agenda for all professors working in cross-cultural settings. PAR provides an approach that will make curriculum internationalization more coherent, just, rational, and informed, as well as more satisfying and sustainable for participants and others involved.

Notes

1. We use the term professor meaning university teacher—someone responsible for university teaching or research supervision.
2. The *Research Quality Framework* and *Learning and Teaching Performance Fund* are major and controversial influences (Thornton, 2008).

References

Altbach, P. G. (1991). Patterns in higher education development. *Prospects, 21*(2), 189–203.
Aulakh, G., Brady, P., Dunwoodie, K., Perry, J., Roff, G., & Steward, M. (1997). *Internationalising the curriculum across RMIT University*. Melbourne: RMIT.
Avis, J. (2005). Beyond performativity: Reflections on activist professionalism and the labour process in further education. *Journal of Education Policy, 20*(2), 209–222.
Biggs, J. (2003). *Teaching for quality learning at university* (2nd ed.). Buckingham, UK: SRHE and Open University Press.
Birrell, R. (2006). Implications of low English standards among overseas students at Australian universities. *People and Place, 14*(4), 53–65.
Boud, D., Cohen, R., & Sampson, J. (Eds.). (2001). *Peer learning in higher education: Learning from and with each other*. St Ives, UK: Clays.
Bretag, T. (2007). The emperor's new clothes: Yes, there is a link between English language competence and academic standards. *People and Place, 15*(1), 13–22.
Brown, R. (2007, July). *Can quality assurance survive the market?* Keynote address to the Australian Universities Quality Forum. Hobart, Australia. Retrieved from http://www.auqa.edu.au/auqf/2007/proceedings/index.htm.
Butorac, A. (1997). *Quality in practice. Internationalising the curriculum in the classroom*. Curtin Western Australia: Curtin University.
Cargill, M., Cadman, K., & McGowan, U. (2001). Postgraduate writing: Using intersecting genres in a collaborative, content-based program. In I. Leki (Ed.), *Academic writing programs. Case studies in TESOL practice series* (pp. 85–96). New York: TESOL.
Carr, W. (2006). Philosophy, methodology and action research. *Journal of Philosophy of Education, 40*(4), 421–435.
Carr, W., & Kemmis, S. (1986). *Becoming critical: Education, knowledge and action research*. London: Falmer.
Fals-Borda, O. (1979). Investigating reality in order to transform it: The Colombian experience. *Dialectical Anthropology, 4*, 33–55.
———. (1989). *Investigating reality in order to change it*. [Videotape]. Calgary: University of Calgary Com/Media.
Freire, P. (1982). Creating alternative research methods: Learning to do it by doing it. In B. Hall, A. Gillette, & R. Tandon (Eds.), *Creating knowledge: A monopoly?* (pp. 29–37). New Delhi: Society for Participatory Research in Asia.
Fullan, M. (1989). *The new meaning of educational change*. London: Cassell.
Giddens, A. (1990). *Consequences of modernity*. Palo Alto: Stanford University Press.
Habermas, J. (1974). *Theory and practice* (Trans. John Viertel). London: Heinemann.
———. (1987). *Knowledge and human interests* (Trans. Jeremy J. Shapiro). Cambridge: Polity.
———. (1996). *Between facts and norms* (Trans. William Rehg). Cambridge: MIT Press.
Haarlov, V. (1997). *National policies for the internationalization of higher education in Europe 1985–2000* (Case: Denmark). Stockholm: Hogskoleverket Studies, Hogskoleverket Agency for Higher Education.
Hadot, P. (1995). *Philosophy as a way of life* (Ed. and Intro. Arnold I. Davidson, Trans. Michael Chase). Oxford: Blackwell.
Ho, D., & Chiu, C. (1994). Component ideas of individualism, collectivism, and social organisation: An application in the study of Chinese culture. In U. Kim, H. Triandis, C. Kagitcibasi, S. Choi, & G. Yoon. (Eds.), *Individualism and collectivism: Theory, method, and applications* (pp. 137–156). Thousand Oaks, CA: Sage.
Hooper, M., Garnett, G., Walsh, L., Vicziany, M., Rizvi, F., & Webb, G. (1999). *Internationalization of the Curriculum*. Melbourne: Monash University (mimeo).
IDP Education Australia (1995). *Curriculum development for internationalization: Australian case studies and stocktake*. Canberra: DEETYA.
Keenoy, T. (2004). Facing inwards and outwards at once: The liminal temporalities of academic performativity. *Time and Society, 14*(2/3), 304–321.

Kember, D. (2000). Misconceptions about the learning approaches, motivation and study practices of Asian students. *Higher Education, 40*(1), 99–121.

Kemmis, S. (2007, October). *Action research as a practice-changing practice.* Opening Address for the IV Congreso Internacional Sobre Investigación-Acción Participativa, University of Valladolid, Spain.

Kemmis, S., & McTaggart, R. (Eds.). (1988). *The action research planner* (3rd ed., substantially revised), Geelong: Deakin University Press.

Kemmis, S., & McTaggart, R. (2000). Participatory action research. In N. Denzin, & Y. Lincoln (Eds.), *Handbook of qualitative research* (2nd ed.) (pp. 567–605). Beverley Hills, CA: Sage.

———. (2005). Participatory action research: Communicative action and the public sphere. In N. Denzin, & Y. Lincoln (Eds.), *Handbook of qualitative research* (3rd ed.) (pp. 559–604). Beverley Hills, CA: Sage.

Kemmis, S., & Smith, T. J. (Eds.). (2008). *Enabling praxis: Challenges for education.* Rotterdam: Sense.

Leask, B. (2001). Bridging the gap: Internationalizing university curricula. *Journal of Studies in International Education, 5*(2), 100–115.

Lindblad, S., & Popkewitz, T. S. (Eds.). (2004). *Educational restructuring: International perspectives on travelling policies.* Greenwich, CT: Information Age.

Lindsay, B. (2008). Breaking university rules: Discipline and indiscipline past and present. *Australian Universities Review, 50*(1), 37–39.

Marshall, J. D. (1999). Performativity: Lyotard and Foucault through Searle and Austin. *Studies in Philosophy and Education, 18*, 309–317.

OECD. (1994). *Education in a New International Setting: Curriculum Development for Internationalization—Guidelines for Country Case Study.* Paris: OECD (CERI).

Pusey, M. (1991). *Economic rationalism in Canberra: A nation-building state changes its mind.* Cambridge: Cambridge University Press.

Ramsden, P. (2003). *Learning to teach in higher education* (2nd ed.). London, UK: Routledge Falmer.

Rapley, J. (2004). *Globalization and inequality: Neoliberalism's downward spiral.* Boulder, CO: Lynne Rienner.

Rizvi, F. (2000). *Internationalization of curriculum.* Melbourne: RMIT University (mimeo).

———. (2004). Globalisation and the dilemmas of Australian higher education. *ACCESS Critical perspectives on communication, culture and policy studies, 23*(2), 33–42.

Schatzki, T. (1996). *Social practices: A Wittgensteinian approach to human activity and the social.* New York: Cambridge University Press.

———. (2002). *The site of the social: A philosophical account of the constitution of social life and change.* University Park: University of Pennsylvania Press.

Swales, J. M. (1990). *Genre analysis: English in academic and research settings.* Cambridge: Cambridge University Press.

———. (2004). *Research genres.* Cambridge: Cambridge University Press.

Swales, J. M., & Feak, C. B. (1994). *Academic writing for graduate students: A course for non-native speakers of English.* Ann Arbor: University of Michigan Press.

Thornton, M. (2008). The retreat from the critical: Social science research in the corporatised university. *Australian Universities Review, 50*(1), 5–10.

Vidovich, L., & Porter, P. (1999). Quality policy in Australian higher education of the 1990s: University perspectives. *Journal of Education Policy, 14*(6), 567–586.

Volet, S. (1999). Learning across cultures: Appropriateness of knowledge transfer. *International Journal of Educational Research, 31*, 625–643.

Watkins, D., & Biggs, J. (Eds.). (1996). *The Chinese learner: Cultural, psychological and contextual influences.* Hong Kong: Central Printing Press.

———. (2001). *Teaching the Chinese learner: Psychological and pedagogical perspectives.* Hong Kong: Central Printing Press.

Weick, K. E. (1976). Educational organizations as loosely coupled systems. *Administrative Science Quarterly, 21*(1), 1–19.

Wenger, E. (1998). *Communities of practice: Learning, meaning, and identity.* Cambridge: Cambridge University Press.

CHAPTER EIGHT

Critical Complexity and Participatory Action Research: Decolonizing "Democratic" Knowledge Production

JOE L. KINCHELOE

In the contemporary globalized knowledge landscape, it is amazing to observe the changes occurring in the nature of research and the politics of knowledge. Much of what is referenced as democratic knowledge and research is anything but democratic as hidden forms of positivism and dominant power insidiously inscribe information work. This chapter explores the dynamics that shape contemporary research and the possibilities for critical practice offered by participatory action research (PAR). As most educational researchers know by now, PAR is collaborative, dedicated to change, and praxiological action. In this critical praxis-based orientation, such research develops and cultivates a reflective community and draws upon literature and theoretical insights from a variety of sources. Such PAR is explicitly political in its orientation as it works to undermine race, class, gender, and sexual oppression and to decolonize diverse forms of knowledge production.

This critical vision of PAR is devoted to a mode of sociopolitical/educational research that is aware of the assumptions that shape its purposes and designs, devoted to the ending of human suffering, focused on consequences of its implementation, and conscious of the ideological and epistemological tenets that inform it. In a mature form of PAR, the categories of participation, action, and research blur. Discrete states of research, participation, and action do not exist. On the contrary, innumerable feedback loops emerge where participants reflect on inquiry and the nature of participation while theoretically informed actions are considered in light of the specific knowledge produced about a set of phenomena. Socioeducational transformation does not take shape at the end of the research but throughout the process of collaboration, inquiry, and action—often in diverse and unexpected ways. To the consternation of many ethics boards and dissertation committees, researchers may modify the objectives and focal points of the research as new insights are gained and/or conditions change (Kincheloe, 2005; Martin, lisahunter, & McLaren, 2006; Wadsworth, 1998).

A central dimension of this chapter involves examining the epistemological and ideological foundations of PAR. Such a task should be a key part of any rigorous PAR, as it alerts participants to the hidden ways the epistemological schema and ideological constructs of dominant power blocs subvert the democratic, anticolonial, and emancipatory dimensions of such critical theoretical/critical pedagogical research (Burbules & Berk, 1999; Kincheloe, 2008). I have watched activist groups, students, and professors undertake PAR with noble intentions—only to fall victim to unexamined epistemological assumptions that reinscribed particular forms of white supremacy, class bias, gender oppression, and colonial relationships.

Social Justice and Scholarly Rigor: Critical Complexity, Epistemology, and the Politics of Knowledge

Those of us in the critical cosmos who question the epistemological and ideological foundations of knowledge have been typically charged with venal scholarly sins. The guardians of Western knowledge often maintain that we are undermining the notion of knowledge itself, subverting the social action that can take place only with the grounding of positive knowledge. Such charges assume the existence of universal criteria for evaluating the "truth-value" of the knowledge anyone produces. The question that critical scholars have asked over the last few decades—how could we have subverted something that never existed?—seems to get to the point. The notion that an objective set of universal criteria for evaluating the truth of propositions ever existed is naïve at best and ethnocentric at worst. Humans, no matter what the time or place are socially constructed, imperfect, and contextualized entities who produce not only fallible perspectives on the world but also fallible criteria for evaluating the validity of such knowledge. Keeping these ideas at the front burners of our consciousness, we become more humble, self-conscious, and insightful knowledge workers.

Thus, embedded in this critical notion of PAR is a new conceptualization of what a social science is and does. To the naïve realists and positivists, such a new social science embodies the epistemological choice of relativism in a reductionistic universe that offers only the possibility of relativism or objectivism in research. To Eurocentric reductionists, the choice of objectivity opens up to the rigorous universal truths of Western science that will maintain the hegemony of Western neocolonial, class elitist, white supremacist, and patriarchal ways of understanding and being in the world. This positivistic objectivism maintains what critical theorist Walter Benjamin (1999) called the dream-filled sleep of the commodity trance of Western consumerism that leads to environmental destruction and human suffering. The choice of subjectivism undermines the privileged status of Western epistemologies and ontologies and catalyzes, as Roger Kimball (1996) phrases it, the descent into tribalism, barbarism, and servitude. Here a positivistic scholarly rigor must be retained to not only save objective truth but also save Western Civilization.

Critical PAR appreciates these power dynamics, these geopolitical dimensions of knowledge production, as it focuses on ways of doing research that promote social justice. The promotion of social justice is not devoted only to the consequences of such research but the means by which it is undertaken. Thus, to the degree that it is possible in a complicated and complex cosmos, PAR carefully attends to dimensions of inclusion and community building in the research process. It is important to discuss the notion of critical complexity used here to epistemologically ground critical PAR: a critical complex epistemology assumes that the mind creates rather than reflects, and the nature of this creation cannot be separated from the surrounding social world. With this notion in mind, many observers have come to the conclusion over the last several decades that the oversimplification of a correspondence epistemology and the dominant forms of knowledge it produces do not meet contemporary social or emancipatory needs.

The social web of reality is composed of too many variables to be considered and controlled in a positivistic model. Scientist Illya Prigogine (1996) labels multiple variables, extraneous perturbations—one extraneous variable in an educational study can produce an expanding, exponential effect. So-called inconsequential entities can have a profound effect in a complex nonlinear universe. The shape of the physical and social world depends on the smallest part. The part, in a sense, is the whole, for via the action of any particular part, the whole in the form of transformative change may be seen. To exclude such considerations is to miss the nature of the interactions that constitute reality. The development of a critical epistemology does not mean that we simplistically reject all empirical science—that would be ridiculous. It does mean, however, that inquirers engaged in critical PAR conceive of such scientific ways of seeing as one perspective on the complex web we refer to as reality.

Reality is too complex and multidimensional to lend itself to fixed views and reductionistic descriptions. Understanding the tendency for reductionism in some of the traditional modes of thinking about curriculum, Kenneth Teitelbaum (2004) maintains that forms of positivism have subverted the effort to gain a more relational perspective on the activity of teaching. In Teitelbaum's estimation, such a relational perspective would connect our understanding of individuals to their social and historical contexts. Teachers' understandings of students in such a conceptualization would be far deeper and helpful in the teaching process. Researchers' understandings would produce modes of knowledge far more helpful to the complex everyday life of the classroom teacher.

Critical educators who take complexity seriously, Stephen Fleury (2004) writes, challenge reductionistic bipolar true or false epistemologies. As critical complex researchers come to recognize the complexity of the lived world with its maze of uncontrollable variables, irrationality, nonlinearity, and unpredictable interaction of wholes and parts, they begin to also see the interpretative dimension of reality. We are bamboozled by a crypto-positivistic science that offers a monological process of making sense of the world. Critical complex scholars who embrace PAR maintain that we must possess and be able to deploy multiple methods of producing knowledge of the world (see Kincheloe & Berry, 2004 on the bricolage). Such methods provide diverse perspectives on similar events and alert us to various relationships between events. In this complex context, we understand that even when we use diverse methods to produce multiple perspectives on the world, different observers will produce different interpretations of what they perceive. Given different values, different ideologies, and different positions in the web of reality, different individuals will interpret what is happening differently. We must understand this complexity to appreciate the complications of gaining knowledge, Charles Bingham (2004) argues. Humans are not atomistic in their ability to acquire knowledge—they must receive help from others to engage in learning.

Bingham's notion of the relationship between knower and known changes the way we approach knowledge, learning, teaching, and research. Critical knowledge work in this complex process is not something employed by solitary individuals operating on their own. Critical scholars use language developed by others, live in specific contexts with particular ways of being and ways of thinking about thinking, have access to some knowledges and not others, and live and operate in a circumstance shaped by distinct dominant ideological perspectives. In its effort to deal with previously neglected complexity, the critical epistemology and the critical PAR it supports appreciate the need to understand these contextual factors and account for them.

Individuals who employ a critical complex epistemology in their work in the world are not isolated, but people who understand the nature of their sociocultural context and their overt and occluded relationships with others. Without understandings of their own contextual embeddedness, individuals are not capable of understanding from where the prejudices and predispositions they bring to the act of meaning making originate. Any critical pedagogy that attempts to deal with the complexity of the lived world must address these contextual dynamics. Patricia Hinchey (2004) illuminates one of the myriad of consequences that occur when the complexities of context are ignored: individuals do not understand the origins of the racial, ethnic, and other forms of prejudice that are unconsciously picked up from their lived contexts. Thus, the transcendence of a reductionistic positivism and passage into a new domain of critical complexity possesses profound consequences.

Many scholars in education and other disciplines have argued that the recognition of complexity in the epistemological domain would undermine our ability to defend the validity of the knowledge we produce because we would have no universal criteria to invoke that were untainted by the context of their production. The knowledge we produced would be useless. Of course, the critical complex epistemological answer to such arguments is that we have *never* had a set of pristine, transcultural/transhistorical epistemological criteria to serve as the final arbiter of truth. A critical complex epistemology frees us from the delusion that such untainted standards exist—a profound contribution to human efforts to understand the world and self. Knowing this, we can operate in a far more humble domain, become more insightful about the forces that shape our own and other people's constructions of reality, gain the ability to understand the dynamics that limit our understandings, appreciate the value of other people's and other cultures' ways of seeing, and discern how to avoid the pitfalls of reductionism.

As we go back to the foundations of Western science in the seventeenth and eighteenth centuries, we see this reductionism manifesting itself in the work of Rene Descartes and Isaac Newton who both saw the world as a giant machine. Newton laid out a universalistic theory of causal determinism, an ultimate mode of reductionism that posited that all motion in the world can be predicted precisely when we know the laws of motion, where a phenomenon is located, and the speed at which it is moving. Thus, it is possible in this framework to predict the future of everything from the largest masses to the smallest objects. One can see that a critical complex epistemology's concern with complexity runs at odds with the Western epistemological tradition. Nature and human behavior do not operate as a machine, for they are both grounded on a complex matrix of interrelationships.

Here rests the nature of *being in the world*. Of course, ontology is the study of being in the world—and the phenomena we study in a critical complex PAR are always ontologically complex, parts of diverse larger processes and contexts. In this complex context a critical PAR is suspicious about dogmatic, reductionistic, one-dimensional definitions of action. Undoubtedly a critical PAR is devoted to action that results in material changes in the well-being of researchers and the communities with whom they have collaborated. However, action can also be defined as the production of transgressive knowledge that helps individuals, who reside far away from the community where the research take place, develop new ways of seeing, new modes of critical consciousness (Harding, 1998; Kincheloe, 2005; Thayer-Bacon, 2000, 2003).

Sometimes the actions that make the most difference in the quality of life of oppressed peoples may surface in unexpected places. Little did Paulo Freire know, for example, when he was engaged in PAR-like research with Brazilian peasants in the 1960s that many individuals in critical pedagogy in North America, Africa, the Islamic World, and indigenous communities around the world would be using the concepts that emerged from such knowledge work in the twenty-first century to bring about diverse forms of social, epistemological, and pedagogical conscientization. PAR needs to include the interrogation of the researchers', the collaborators', and the reflective community members' contextualized understandings of the research project and the construction of their consciousness in general.

Too often PAR glorifies the perspectives of the collaborators and community members in a form of left democratic essentialism. Essentialistic tendencies must be questioned in critical PAR—a questioning that allows for a rigorous and genuine dialogue between researchers coming from diverse places in the web of reality. Essentialism constitutes a fetishization of democratic inclusivity that undermines theorizing and action that understands the sociopolitical construction of all perceptions. This is not a popular topic to address in contemporary discussions of PAR and may be misrepresented as a denial of the democratic impulse—I hope that advocates of PAR will view this as an opportunity to open new enhanced forms of dialogue and more informed modes of democratic inclusivity. We are attempting to move to a more informed understanding of PAR and what it can accomplish in the sociocultural, political, and pedagogical domains.

New Venues, Participants, and Ways of Perceiving in Research

A principal aspect of a critical PAR involves helping formulate the long overdue moment when diverse societies talk back to the authoritarian scientific establishment. Critical theoretical insights into both the contextualization and the archeological analysis of the ways final truths and objectivity are constructed have the potential to radically change the nature of research. The defenders of the status quo recognize this possibility and over the last couple of decades have worked tirelessly to make sure that this does not happen. The critical power of contextualization grants new insight into the process of construction of "unassailable," universal notions of dominant Western forms of research. In the constructions of neopositivist research such contextual dynamics are another effort to destroy the superior work of Euroscience. The rigorous hermeneutic act of contextualization in this neo- and often crypto-positivist context is viewed as an attack on objectivity. Cultural workers and educators who employ PAR must understand these concepts and work to be good hermeneuts as well as good activists.

Criticalists who make use of PAR are particularly interested in the questions: Whose view of the world do researchers give to their readers?

How do epistemological and ideological choices shape the view of the world that is disseminated? To answer such questions, advocates of PAR must rigorously understand and distinguish between the competing claims about the nature of reality (ontology) and sociocultural relationships (ideology). How do these claims work to include or exclude differing race, class, gender, and sexual groups? If researchers are unable to answer these questions, they may have the intention of performing democratic research but not the theoretical understanding necessary to accomplishing the task.

Without the aforementioned forms of hermeneutic and ideological analysis, researchers find it hard to escape from the tacit inscriptions of what Sandra Harding (1998) calls the internalist epistemology of Western science. Such a position asserts that the success of Western research rests on its internal structures that always operate to maximize objectivity and rationality. There is only one legitimate science, positivists posit, and this objectivistic, rationalistic variety is *it*. This view persists even after four decades of historical, sociological, discursive, epistemological, and ethnographic studies have documented the impact of social, paradigmatic, and political economic influences on objective science. With the efforts to recover various forms of dominant power over this same period, the attempt to reestablish the superiority of naïve realism and positivism continues unabated at the end of the first decade of the twenty-first century. To hell with participation and action in research, we have to preserve Western supremacy (Harding, 1998; Nowotny, 2000; Steinberg & Kincheloe, 2006).

Critical PAR can only exist in conflict with scientific neopositivistic research with their profoundly different assumptions about knowledge and its production. In distinction to critical PAR, scientific positivism maintains that

- All observers receive the same information about phenomena. *Advocates of critical PAR maintain that data uncovered in diverse contexts by observers from divergent backgrounds will be viewed differently. There is always room for dissimilar interpretations of the meaning of phenomena.*
- Rigorous researchers produce observation statements about the world that support universal theoretical constructs. *Advocates of critical PAR construct compelling interpretations of data that provides insights on which actions can be formulated.*
- Good researchers produce knowledge that is beyond interpretation. If it is produced correctly, it will provide one unambiguous and univocal meaning. *Given the complexities of context, interpretation, and polysemy of language, advocates of critical PAR understand there is no final meaning of the knowledge they produce. Different historical times and places will induce observers to view such knowledge in unexpected ways.*
- Researchers operating in the proper way produce knowledge that can be quickly falsified or confirmed in relation to larger universal

theory. *Advocates of PAR understand that knowledge always exists within an interpretative framework that changes as the hermeneutic matrix is modified.*
- Good research makes sure that human inscriptions are removed from the knowledge produced. *Advocates of critical PAR understand that the knowledge they produce is always related in overt and covert ways to the needs, interests, and welfare of all.*
- Serious research always views phenomena in question as examples of a general type. *Advocates of critical PAR are interested in both the general and particular nature of any phenomenon such as an individual lived experience or a particular sociocultural/educational circumstance.*
- Rigorous inquiry is produced anonymously—it is irrelevant who produces knowledge for anyone given the same circumstances and following correct procedure would discover the same information. *Advocates of critical PAR appreciate the significance of who produces the knowledge at hand. The consciousness, sociocultural backgrounds, goals, and epistemological and ideological predispositions of the researcher all inform the process and product of research* (Kemmis & McTaggart, 2005; Rouse, 1987; Steinberg & Kincheloe, 1998).

When researchers ignore the preceding points and work toward producing neutral, objective knowledge, they fall into the irrational rationality trap of neopositivistic research. Such a quest invariably ends up with the researcher unconsciously focusing on what such an epistemology and the research designs and methods it supports allow it to address rather than on what holds the most importance for human affairs. Those individuals affected by such research simply have no voice in this positivist configuration. Any protest they might direct at the researchers is not viewed as important input to the larger knowledge production process but as the uninformed protestations of special interest groups. The expression of such concerns is not simply an interruption of the scientific process but a force that threatens the objectivity of the scientific enterprise. In this context, one begins to understand the threat to positivist orthodoxy presented by critical PAR.

The Normalizing Politics of Epistemological Universalism

Universal truths about human behavior and social dynamics are frightening. In the eyes of criticalists, they are the Freddies and Jasons of epistemology—*Nightmare on Positivist Street*. Universalism is a key part of rationalization in that human beings via the gospel of science are granted to ability to discover all of the laws of the cosmos and subsequently subsume nature to *man's* needs—especially *his* quest for political domination and economic development. Science provided Westerners the power to conquer nature a la Descartes's and Bacon's dreams. On the colonial landscape, psychologists, social and physical scientists, economists, and educators come together to produce the structural conditions and human capital that make possible good business climates, the growth of capital, the maximization of profits in every corner of the planet.

For many criticalists from around the world, positivist universalism reveals itself as a dimension of dominant power that perpetuates oppression and silences voices outside the mainstream. In this power-driven context, assertions of universalist knowledge invalidate the expertise of the poor, racially marginalized groups, indigenous peoples, women, and colonized cultures that has allowed them to not only survive in adverse sociocultural and political circumstances but often interact in sustainable ways with the natural world. Here in its dismissal of diverse forms of human genius, positivist universalism again reveals the irrationality of Western rationalism. Ignorance in the guise of learned expertise prevails. Those colonized and exploited human beings who are treated like stray dogs do not need the epistemological scraps of Western science. Critical PAR values their ways of seeing and being.

All knowledge—positivist universalist knowledge included—is a local form of information. Positivist scientific knowledge involves knowing about the overt and tacit knowledges of a particular community of human beings—scholars typically housed in universities. Such knowledge also entails cognizance of the cultural capital of this group as well as the tacit rules and protocols members unconsciously embrace in their interactions. It also involves knowing how to handle oneself around the laboratory, the field site, or the interview setting. These are local knowledges that are discursively, historically, and culturally specific. The discursive strategies, mores, folkways of the scientific site vary from culture to culture and from historical era to historical era. Claims to universality fall short as we examine the specifics of diverse scientific enterprises. The subjectivity of the researcher is inseparable from the context of her interactions. The ethnocentric logic embedded in universalist impulses emerge in the assertions that it is Western modes of producing knowledge that are universalized—not African or Asian modes of knowledge work.

Critical PAR maintains as long as researchers fail to understand these epistemological, cultural, and ideological dynamics, there is little need for overt forms of censorship. These tacit epistemological, cultural, and ideological forces keep Western knowledge safe for human consumption. Watch the news on ABC (Disney), read Rupert Murdock's newspapers, examine many dissertations emerging from contemporary universities, peruse the academic research studies funded by Pfizer Pharmaceuticals, explore the educational studies employed to justify school reform in many Western nations—dominant power wielders are not fearful of what such data might suggest. We have entered a new era of corporatized knowledge where public information becomes more limited to what serves the needs of various power blocs; where even academic knowledge is tamed and recast to fit such elite demands.

As Western power wielders have engaged in preemptive wars, continued to redistribute wealth from the poorest peoples in the world to the richest, undermine public education, and change the climate of the planet, knowledge domains such as television and radio have become corporatized and their reporting trivialized. Following the dictates of the holy market, coverage of starlets in rehab increases while coverage of the aforementioned issues decreases. The corruption of Western governments and the information they produce helps corporations' political pawns win great victories—either through the effects of the corruption itself or via the electorate's cynicism and disengagement from politics that results from such government sleaze. Although individuals living in these societies continue to do amazing things and develop brilliant forms of expertise, their political lives as citizens of democratic states are truncated by this neocolonial, hyperreal politics of knowledge and crypto-positivist epistemology.

This is what makes a critical PAR such a dangerous activity. Those who would operate to identify this oppressive politics of knowledge, to engage in modes of decolonizing activism are vulnerable to punishment. Twenty-first century sociopolitical and pedagogical heretics find their professional aspirations undermined; in the academy tenure is threatened. Unless critical academics are united and committed to the larger social good and the welfare of their brothers and sisters in education and other domains, the day is coming when efforts to produce critical PAR will be swiftly quashed by colleagues and administrators serving the hegemonic needs of the empire.

Michel Foucault (1980) wrote of the need for dominant power to construct a disciplined and well-regulated workforce—a congregation of docile bodies. Even in the production of information, rigorous knowledge becomes that which follows the proper modus operandi prescribed by dominant power. If such epistemological procedures are not respected, then the data constructed possess no worth. Critical PAR is directly focused on disrupting these modes of epistemological and ideological oppression. Only when researchers understand the sociocultural, political, historical, and philosophical issues surrounding the production of knowledge will practical antihegemonic information find its ways into the hands of the public (Castro-Gomez, 1998; Harding, 1998; Jardine, 2005; Rouse, 1987; Saul, 1995).

What Donaldo Macedo (2006) calls a literacy of power is central in a critical complex PAR. Such a critical literacy changes our understanding of the relationship between knowledge and power and allows us to view the fingerprints of power on information long held as sacred by many. Issues that many have viewed only as the concerns of philosophy or research begin in a critical PAR to be looked at as political questions of power relations. We begin to understand that discourses of power help shape how and what we think about a particular topic. Our dangerous PAR uncovers these discourses and discursive practices and the insidious ways they operate to oppress the dispossessed and maintain the status quo. The expose of discursive specificity directly challenges the universalism of positivism. Such a challenge constitutes the heart and soul of a complex criticality, as it exposes both sociocultural and political economic factors in the construction of colonialism and other forms of hegemony.

"True science," crypto-positivists maintain, could not exert any political force. Such positivists ask: How could the simple production of truth be seen as a politicized process? This universalism and political neutrality form the grounding for both old overt and new covert forms of positivism. And such positivism pushes us to the brink of geopolitical, social, environmental, educational, and other modes of cataclysm. As we teeter on the cusp of catastrophe, human beings must make some difficult choices. We must find new ways to deal with the mess that we have made in innumerable domains. Critical PAR understands the epistemological and ideological crises of Western research, as hegemonic scholars from North America and Europe battle to retain their exalted position in particular and Western supremacy in general. Skepticism about such superiority is emerging from diverse locations. Hegemonic forms of positivism can keep the lid on such disbelief and the anger surrounding it for only so long (Bettis & Gregson, 2001; Harding, 1998; McLaren & Kincheloe, 2007).

White Supremacy and Epistemology: Decolonizing Research

In the name of rigorous research and modes of scientific realism, many ways of perceiving the world by many women, poor people, and particular "ethnic" groups are dismissed as inferior. One has to be very careful in this epistemological domain to avoid essentialism of any variety. Women operate in a variety of epistemological modalities, as do poor people, and peoples from diverse cultures. This critical theoretical point involves the *tendency* for positivism to reflect dominant masculine, upper-middle/upper class, and Western perspectives. This hegemonic tendency works to support forms of colonialism and neocolonialism in the epistemological and political economic spheres. Thus, white supremacy, patriarchy, class elitism, and colonialism create an epistemological spigot that shuts off the flow of information from individuals and groups that fall outside the dominant community. At the base of such a reality rests the foundations of critical PAR. Since such exclusionary practices—such a regressive politics of knowledge—exist, then supporters of good research and social justice must develop forms of practice that challenge such oppressive ways of operating.

In this context Western culture slowly sinks into a form of social senility, as it excludes viewpoints that challenge its unexamined assumptions, its sense of superiority. The Western casserole of ordinary consciousness, of normal(izing) science bloats us with affirmations of our own intellectual preeminence and our moral righteousness in the effort to maintain neo-colonial relations. A critical PAR has to constantly fight to make these power-related dynamics known, to question what passes as a *democratic* politics of knowledge/epistemology. Such struggle is necessary because positivism claims to present a value-free, correspondence epistemology. Rigorous science, positivists maintain, follows the rules of the scientific method and produces data that *reflects* (corresponds to) the reality that is out there. Although some researchers admit that the attainment of mirror-like flawlessness is quite difficult, most believe that the attempt to provide such precision in correspondence is the true objective of scientific inquiry.

Such a reductionistic form of realism is an important manifestation of dominant power as it insidiously positions itself as not only value-free but also culture-free. Where a researcher resides in the web of reality, therefore, in the positivist cosmos, has nothing to do with what he or she perceives. When advocates of a critical PAR refer to the value of standpoint epistemologies that come from individuals' particular vantage points on the world, they are speaking in a tongue incomprehensible to the epistemological pirates of positivism. No matter where we stand in the web of reality, if we use the "proper" research methods we should all perceive the same phenomena in the same way. In such a reductionistic framework of knowledge PAR is viewed as a gigantic waste of the researcher's time, for culture and politics have nothing to do with research. Such perspectives reflect the zenith of Western scientistic modes of decontextualization, reductionism, and abstraction.

The ways researchers make decisions about which phenomena to investigate, how they assess the consequence of the information encountered, and the means by which they ascribe particular meanings to a body of data are always in part shaped by the social, cultural, geographical, discursive, epistemological, ideological, and research practices with which they are familiar. When sociologists, historians, and philosophers of science have studied research practices over the last three or four decades, they find that one of the most important dimensions of scientific work has involved the effort to *protect* research from these types of *contaminating* influences. Such dynamics and a reflective understanding of the myriad of ways they influence our understandings of education and other phenomena are powerful dimensions in improving the way we conduct research. A critical mode of PAR sidesteps the pitfalls of a positivistic universalism and engages multiple forms of diversity in its pursuit of knowledge that leads to praxis (Harding, 1998; McClure, 2000; Semali & Kincheloe, 1999).

PAR and the Power of Subjugated Knowledge: The Insights Derived from Difference

Knowledges constructed by colonized peoples and indigenous peoples around the planet reflect what might be called the colonial divergence;

that is the impact colonialism made on seeing the world and being in the world. Peoples who have been subjugated and classified as inferior by colonial, political, economic, and epistemological systems of classification will produce different knowledges that those who come from societies implicated in the colonization process. Again, employing a critical complex epistemology advocates of a critical PAR listen carefully to what those subjugated in this ever-evolving colonial system have to tell us. In the spirit of PAR they become the co-researchers of the oppressed (Freire, 1970, 1997). Historically, when Westerners have addressed epistemology and knowledge production (not to mention curriculum development), this colonial divergence has been erased. Aware of the power of the colonial divergence, advocates of a critical PAR can begin to decolonize the knowledges they encounter/produce.

Oppressed and colonized peoples have every right to feel suspicious and resistant when they hear or read some Western researcher advocating their inclusion via PAR in the research act. Why would Western researchers know how to produce knowledge that would be useful to those whose history has been marked by Western colonial exploitation of their resources and/or labor? A critically informed PAR is comfortable with collaboration with ethnographic teams constructed by individuals from diverse social locations, interpretive communities that include members with profoundly different relationships to colonialism and subjugation.

Diverse collaboration can not only produce knowledges that are respectful and relevant to oppressed groups but also engage in more sophisticated and ontologically powerful modes of critique of data constructed by dominant power. Such critical collaboration moves researchers to new levels of understanding the complex relationship between theory and practice. We are better able as inquirers to free ourselves from the oppressive research practices in which we may have been engaged and to gain deeper insight into our role as emancipatory agents of history. Without a critical epistemology and transformative politics of knowledge, the story of the colonized will continue to be told by the colonizers. Reclaiming these stories and this history is a central task of a critical PAR and for the general process of decolonialization.

Western positivistic epistemology and the hegemonic knowledge it has produced involve a way of seeing that has emerged in collusion with colonial perspectives on the subjugated and the needs of the empire. A critical PAR cannot be allowed to be simply another epistemological position that merely replaces the previous exploitative mode of research with a new but still hegemonic mode of understanding knowledge production. This oppressive historical procession must come to an end. Previous marchers in the parade excluded the oppressed from participation in the research process. Too many contemporary advocates of PAR have failed to ask hard questions about the nature of participation. Without such complex and intense questions, PAR too often migrates to one of two positions: a research method/design that (1) romanticizes and essentializes the perspectives of the oppressed and fails to question the diversity of viewpoints among subjugated groups; (2) embraces facile notions of participation that serve as new and more hegemonically sophisticated modes of exclusion.

The participatory and action dimensions of critical PAR are complex. Participation does not mean that all parties have to agree on all aspects of a research project. Different individuals in the research community may take differing forms of action as a result of the research project (Harding, 1998; Jardine, 2005; Martin, lisahunter, & McLaren, 2006; Parker, 1999; Smith, 1999; Wadsworth, 1998). Critical PAR opens an ongoing dialogue among a variety of participants and stakeholders that hopefully will continue long after more traditional researchers believe the research has ended. Moving from what is to what could be is a necessary part of a critical PAR, but what constitutes the new and improved condition is never obvious to all parties. No one ever said making the future was easy.

References

Benjamin, W. (1999). *The arcades project*. (H. Liland, & K. McLaughlin, Trans.). Cambridge, MA: Harvard University Press.

Bettis, P., & Gregson, J. (2001). *The why of research: Paradigmatic and pragmatic considerations*. Retrieved from http://www.csulb.edu/colleges/chhs/departments/professionalstudies/documents/Chapter_1.doc.

Bingham, C. (2004). Knowledge acquisition. In J. Kincheloe, & D. Weil (Eds.), *Critical thinking and learning: An encyclopedia for parents and teachers*. Westport, CT: Greenwood.

Burbules, N., & Beck, R. (1999). Critical thinking and critical pedagogy: Relations, differences, and limits. In T. Popkewitz, & L. Fendler (Eds.), *Critical theories in education* (pp. 25–43). NY: Routledge.

Castro-Gomez, S. (1998). Traditional and critical theory of culture: Postcolonialism as a critical theory of globalized society. Retrieved from http://www.javeriana.edu.co/pensar/Sc5.html

Fleury, S. (2004). Critical consciousness through a critical constructivist pedagogy. In J. Kincheloe, & D. Weil (Eds.), *Critical thinking and learning: An encyclopedia for parents and teachers*. Westport, CT: Greenwood.

Foucault, M. (1980). *Power/knowledge: Selected interviews and other writings, 1972–1977*. NY: Pantheon.

Freire, P. (1970). *Pedagogy of the oppressed*. New York: Herder & Herder.

———. (1997). *Pedagogy of the heart*. New York: Continuum.

Harding, S. (1998). *Is science multicultural? Postcolonialisms, feminisms, and epistemologies*. Bloomington: Indiana University Press.

Hinchey, P. (2004). Diversity and critical thinking. In J. Kincheloe, & D. Weil (Eds.), *Critical thinking and learning: An encyclopedia for parents and teachers*. Westport, CT: Greenwood.

Kemmis, S., & McTaggart, R. (2005). Participatory action research: Communicative action and the public sphere. In N. Denzin, & Y. Lincoln (Eds.), *The Sage handbook of qualitative research* (3rd ed.). Thousand Oaks, CA: Sage.

Kimball, R. (1996). The killing of history: Why relativism is wrong. *The New Criterion, 15*(1). Retrieved from http://www.newcriterion.com/articles.cfm/killingofhistory-kimball-3484

Kincheloe, J. (2005). *Critical constructivism*. New York: Peter Lang.

———. (2008). *Critical pedagogy* (2nd ed.). New York: Peter Lang.

Kincheloe, J., & Berry, K. (2004). *Rigour and complexity in qualitative research: Conceptualizing the bricolage*. London: Open University Press.

McClure, M. (2000). Chaos and feminism—a complex dynamic: Parallels between feminist philosophy of science and chaos theory. Retrieved from http://www.pamij.com/feminism.html
McLaren, P., & Kincheloe, J. (Eds.). (2007). *Critical pedagogy: Where are we now?* New York: Peter Lang.
Macedo, D. (2006). *Literacies of power: What Americans are not allowed to know* (2nd ed.). Boulder, CO: Westview.
Martin, G., lisahunter, & McLaren, P. (2006). Participatory activist research (teams)/action research. In K. Tobin, & J. Kincheloe (Eds.), *Doing educational research: A handbook*. Rotterdam, the Netherlands: Sense Publishers.
Nowotny, H. (2000). Re-thinking science: From reliable knowledge to socially robust knowledge. In H. Nowotny, & M. Weiss (Eds.), *Jahrbuch 2000 des Collegium Helveticum*. Zurich, Switzerland: ETH.
Parker, L. (1999). Historiography for the new millennium: Adventures in accounting and management. *Accounting History*. Retrieved from http://www.findarticles.com/p/articles/mi_qa3933/is_199911/ai_n8865215/pg_6
Prigogine, I. (1996). *The end of certainty: Time, chaos, and the new laws of nature*. New York: Free Press.
Rouse, J. (1987). *Knowledge and power: Toward a political philosophy of science*. Ithaca, NY: Cornell University Press.
Saul, J. (1995). *The unconscious civilization*. New York: Simon & Schuster.
Semali, L., & Kincheloe J. (Eds.). (1999). *What is indigenous knowledge? Voices from the academy*. New York: Garland.
Smith, L. (1999). *Decolonizing methodologies: Research and indigenous people*. New York: Zed.
Steinberg, S., & Kincheloe, J. (Eds.). (1998). *Students as researchers: Creating classrooms that matter*. London: Falmer Press.
———. (2006). *What you don't know about schools*. New York: Palgrave.
Teitelbaum, K. (2004). Curriculum theorizing. In J. Kincheloe, & D. Weil (Eds.), *Critical thinking and learning: An encyclopedia for parents and teachers*. Westport, CT: Greenwood.
Thayer-Bacon, B. (2000). *Transforming critical thinking: Thinking constructively*. New York: Teachers College Press.
———. (2003). *Relational "(e)pistemologies."* New York: Peter Lang.
Wadsworth, Y. (1998). What is participatory action research? *Action Research International*. Retrieved from www.scu.edu.au/schools/gcm/ar/ari/p-ywadsworth98.html

CHAPTER NINE

Reconceptualizing Participatory Action Research for Sustainability Education

ELIZABETH A. LANGE

> True unity in the individual and between [humans] and nature, as well as between [human] and [human], can arise only in a form of action that does not attempt to fragment the whole of reality... *To be confused about what is different and what is not, is to be confused about everything.* Thus, it is not an accident that our fragmentary form of thought is leading to such a widespread range of crises, social, political, economic, ecological, psychological, etc., in the individual and in society as a whole.
>
> —Bohm, 1980, p. 16; emphasis in original; inclusive language added.

Introduction: Seeking PAR Elasticity

Given the simultaneous emergence of action research on several continents and in several disciplines, it could be said that there are at least three strands of Participatory Action Research (PAR)—originating in the international development field, the education field, and the community development field. In each case, the purpose of PAR has been one of disrupting hegemony, whether the hegemony of positivist social science over popular knowledge, the hegemony of social engineering projects over community participation, or the hegemony of academic knowledge over professional practice. However, in considering how to engage PAR as a process for educating Canadian adults about sustainability, no existing PAR approach coheres well with sustainability theory. Therefore, this chapter details a theoretical reconceptualization of PAR by first illustrating the integration of these three strands of PAR with sustainability theory. Second, new ontological, epistemological, pedagogical, and social change assumptions emanating from sustainability theory are described, as part of this reconceptualization. Third, the chapter details how this reconceptualization of PAR can be enacted as part of sustainability education, including the impact in terms of socioecological change. Overall, this chapter attempts to disrupt the modernist fragmentary thought forms within existing action research theory and proposes a reconceptualization of PAR as a critical living practice—no longer as a tool but rather as a "set of relations among persons, their histories, their current situations, their dreams, their fantasies, their desires" (Carson & Sumara, 1997, p. xx). In this conceptualization, PAR is not instrumentalized but is a moving, elastic, living form that is continuously emerging, as participants interrogate and transform unjust and unsustainable relations and the societal structures in which they are embedded.

Strands of PAR: Disrupting Hegemony

The following offers a very brief synopsis of each of the three historical strands of PAR, identifying elements vital to a reconceptualization of PAR.

PAR and International Development

In the global South, it was the failure of northern-inspired development approaches and the limitations of the positivist research paradigm that led to the desire to create knowledge, educational practices, and social action for transforming inherently alienating and unjust social structures (de Souza, 1988). The task was to develop social science methods that could break the cultural and epistemological dependency of poor communities in Southern nations (Hall & Kassam, 1985). As Budd Hall (2005) tells the story, the term *participatory research* (PR) began in Tanzania as a description for community-based approaches to knowledge creation, which merged the processes of social investigation, education, and action. In Latin America, *action research* also addressed issues regarding who should benefit from research, what the aims of research should be, who should be involved, and what methods would best contribute to societal transformation toward social justice. Hall (2005) suggests it was Orlando Fals-Borda from Colombia who coined the phrase participatory action research or PAR to express the commitment of activist-scholars to use their intellectual and organizing skills to strengthen social movements for liberatory social change. Thus, participatory research is a research tool to enable the popular classes (working class and peasants) to identify their issues and generate "popular" knowledge useful for discerning effective actions for transforming society toward social justice.

Pablo Latapi (1988) asserts that PAR is also tightly connected to the popular education methodology of Paulo Freire (1970). Freire's pedagogy of the oppressed clearly specifies both a political commitment to the exploited and oppressed and a coherence with an ecopedagogy (Gadotti, 2000). Thus, PAR is appropriate for implementing a Freirean-informed sustainability pedagogy that expresses a political commitment to solidarity with the marginalized and for building links to social movements, as harbingers of social change.

PAR and Community Social Action

In the North, contributions to the practice of PAR came from the Action Sociology of the Frankfurt School, which questioned conventional social research and was concerned with redefining and remaking knowledge (Hall, 1981; Latapi, 1988). Through fostering civil dialogue and social analysis and guiding a community in social action, action research assists in the preservation of a vibrant democracy (Carr & Kemmis, 1986). However, with information more readily accessible in the global North, the task among Northern grassroots groups has been to reappropriate, not necessarily produce, knowledge for their own needs, thus democratizing knowledge access (Gaventa, 1988). As Gaventa (1988) summarized from his experiences at the Highlander Center in Appalachia:

> People may discover for themselves dominant knowledge or interpretations of reality which do not conform to their own experience... Or the process of popular investigation may reveal previously hidden information that does confirm through "official" knowledge what the people have suspected from their own experience. (p. 21)

As people develop techniques for information gathering, the empowerment they feel often can unleash a social movement around an issue, whether civil rights, environmental contamination, or manipulative labour practices.

Thus, an important aspect of PAR in both hemispheres is its collective nature, both in learning and action. Yet, one of the most significant differences between these two strands is that international development PAR is often centered on geographical communities, whereas community social action PAR generates communities of interest, around a shared social identity or issue of concern. Thus, another vital element for reconceptualizing PAR is engaging a community of interest in collectively analyzing and organizing for social change, ensuring that the knowledge benefits those involved in its reappropriation.

Educational Action Research

Carr and Kemmis (1986) assert that PAR is the "research method of preference whenever a *social practice* is the focus of research activity" (p. 165; italics in original). Somekh (2006) adds that whenever research is a systematic intervention in a social practice, carried out in partnership with participants, this constitutes action research. This definition led to the third strand of PAR—educational action research or more generally, profession-based action research. When the reflective practitioner movement (Schön, 1983) intersected with the teacher-as-researcher movement, it significantly impacted the field of teacher professional development. In the United Kingdom, John Elliott (1991) asserted that the fundamental aim of action research should be to improve practice, not to produce knowledge. He advocated moving away from the theory-driven, technical rationality of action research toward an ethically based action research, that he conceived of as a "moral science" (p. 52). He suggested that the redeeming quality of action research should be the capacity to build collaborative reflection and foster practical wisdom among practitioners. The Australian action research movement went a step further to demonstrate the possibility of action research for the structural transformation of schools (Carr & Kemmis, 1986). For them, issues that arise in an individual's teaching practice are social matters that require collective dialogue, theorizing, and action, not just the reflective practices and practical judgement of one teacher within one classroom. In Canada, Carson (1993) followed Elliott's notion of action research as an ethical activity seeing great potential for a postmodern practice of action research for teacher practitioners. Carson and Sumara (1997) conceived of action research as a living practice that attends to the way the investigation and investigator coemerge, the complexity of relations and identities, and how modernist concepts shape the normalizing practices of institutions and structures. It is a research practice of deep awareness of being—not a product, tool, or method. As they suggest, action research as living practice occurs when "who one *is* becomes completely caught up with what one knows and does" (p. xvii; italics in original). This conception of action research was important for studying my own PAR practice but also for its theoretical coherence with sustainability theory.

Overall, the elements of PAR necessary to a reconceptualization for the purposes of sustainability education include a commitment to solidarity with the marginalized, links with social movements, engagement of communities of interest, reappropriation of knowledge, and PAR as a way of knowing and being. Sustainability theory, however, adds some new premises for a reconceptualization of PAR.

Sustainability Theory and PAR

Environmental discourse links political, economic, and political thought into specific paradigms, described by Merchant (2005) as Deep Ecology, Spiritual Ecology, Social Ecology, Environmental Justice, Ecofeminism, Sustainable Development, and Sustainability. While the term sustainability is often co-opted, I have defined a sustainable society as one that satisfies its needs without diminishing the prospects for health and justice for self, other peoples, future generations, or the environment. I adapted Lester Brown's definition (1981) to holistically incorporate personal and community sustainability as well as social justice with environmental sustainability. The following contrasts Sustainable Development with Sustainability and identifies key premises for informing a new practice of PAR.

The Brundtland Commission Report, *Our Common Future* (Brundtland, 1987), introduced the concept of sustainable development (SD), a paradigm embedded in mainstream environmentalism with the goal of making existing production systems more ecologically sustainable. This "greening" discourse seeks to reconcile environmentalists and corporate business by seeking a greening of existing business while maintaining rapid economic growth, increasing consumption, and technological innovation. Although some of the policy adjustments, technological fixes, market solutions, and managerial processes it advocates might be important short-term initiatives, this paradigm is criticized for not addressing the structural roots of poverty and environmental degradation, for not considering social injustice and ecological injustice as manifestations of one system of domination and exploitation, for assuming that conventional economic growth can be harmonized to ecosystem limits, and for maintaining the anthropocentric, resourcist, developmentalist, social engineering paradigm.

Edwards (2005) considers the Sustainability Revolution more sweeping than the eighteenth century Industrial Revolution, contrasting Sustainability with Sustainable Development. He identifies its five characteristics: a large number of grassroots groups prevalent globally, uncentralized multiple nodes of independent activity that focus on diverse issues but with remarkably similar values and intentions, leadership by decentralized visionaries, and varying modes of action, including oppositional and alternative. Hawken (2007) suggests that historically, this is likely the largest movement ever and he estimates there are at least thirty thousand sustainability groups in the United States and tens of thousands of groups worldwide.

The Sustainability (or Sustainable Livelihoods/Sustainable Communities) approach is based on principles such as localization and bioregionalism where self-sufficient communities meet their own needs largely from within their own bioregion through biomimicry that uses locally appropriate technology mimicking the synergy and wisdom inherent in natural systems enabling survival over millennia. Restoration ecology restores human-disturbed ecosystems and implements no waste and renewable energy principles. Indigenous sustainability preserves traditional lands and ways of knowing. Other sustainability principles include restorative economics where commerce is redesigned to give back to the natural world as much as is taken, participatory democracy where decision making responds democratically to local needs while keeping the global in mind, and redefining wealth as social and spiritual wealth rather than material goods. In terms of social change, it integrates both cultural transformation, as advocated by deep ecologists, alongside economic and political transformation, as advocated by social ecologists. The Sustainability approach avoids expertism preferring to engage grassroots populations in rethinking and reconstituting how we heal, teach, work, and provide food and shelter in our societies. This approach is meant to empower civil society to embrace life-giving values that can inform the creation of just social structures and processes—where interactions harmonize with the biological, life-preserving systems of Earth (Korten, 2006).

Sustainability theory draws its premises from living systems theory, based on the New Science that incorporates relativity theory, quantum mechanics, process physics, complexity theory, and Gaia theory (Capra, 1996). Early twentieth-century discoveries by Einstein, Bohr, and other physicists, demonstrated that the existentials of form, matter, time, and space are not fixed and linear but are dynamic, curved, and flowing. Subatomic particles are not things, but interconnections. Matter and energy are interchangeable, either as particles or waves, often emergent with the act of observation. From these insights, David Bohm (1980) proposed a new model of reality where the universe is an undivided wholeness enfolded into an infinite, timeless background source, continually unfolding into visible and temporal material and then, as part of the continual flow of energy, returning to the undivided wholeness. While conscious thought can grasp the unfolded, only a consciousness beyond rational thought can experience the enfolded, the implicate order. Further, through their research into mind, cognition, and consciousness, Maturana and Varela (1980) suggest that our world is continuously shaped by our actions as well as by all the natural processes around us. In other words, mind and world arise together (Varela, Thompson, & Rosch, 1991). They propose that there is still a real material world, but it does not take shape independent of cognition. Cognition is not a representation of an independently existing world, but a "continual bringing forth of *a* world through the process of living" (Capra, 1996, p. 267). Through the notion of embodied action, they theoretically negotiate between realism (a pregiven outer world) and idealism (the projection of a pregiven inner world) as a way to end the fragmenting illusion of a separate, independent self and a world out there. Thus, sustainability and living systems theory pose significant challenges to existing conceptualizations of PAR and demand new ontological, epistemological, pedagogical, and social change premises for use within a process of sustainability education.

PAR and Ontological Assumptions

Spretnak (1999) considers ecological postmodernism to go beyond the narrow confines of environmentalism and beyond the groundlessness of postmodernism to reconstruct a grounded, deeply ecological, and spiritual ontology. She asserts it is possible to retain the realism implicit in the critical PAR tradition while acknowledging the insights of an ecological postmodernism where the dynamic unfolding nature of reality can never be fully captured in any theory or narrative, particularly the nonlinearity, interrelatedness, multiplicity, and multidimensionality of life. She insists this is not the creation of another essentializing or totalizing narrative, but a storytelling that honors the mysteriousness of a reality that is implicit and not fully describable, specifiable, or representable. Such an ontology searches for nonfragmentary both/and representations. Paradox rather than contradiction assists in understanding reality as real *and* constructed, the cosmos as unity *and* diversity and, as I discuss below, knowledge as whole *and* partial, thinking as reason *and* imagination, and adult development as autonomy *and* relatedness.

PAR and Epistemological Assumptions

Drawing from community social action PAR, it was important in my practice to democratize access to knowledge in a Northern context—especially knowledge that never surfaces into the mainstream media but is hidden or suppressed or eclipsed by a sensationalist press. Yet, the dominant sociological analysis is that the middle class is in a moral malaise due to narcissistic individualism, that they have deliberately retreated from a vigorous dialogical political culture, and that they are in denial about the urgency of the social and environmental crises facing humanity. The adults I had contact with through my university extension classes did express a desire for personal transformation, particularly given the impacts of neoliberal globalization on their work lives, and they also expressed the desire for larger social change related to ecological issues. The challenge was to catalyze these desires into the formation of a community of interest engaged in analyzing Western societal structures and lifestyles and transforming them as part of socioenvironmental change. Could middle-class individuals with little politicization be shifted out of a passive consumptive-spectator mode into an active mode, where they seek out or generate the knowledge they need as part of their empowerment and civic engagement? Rather than considering them ideologically duped or fundamentally self-interested, PAR offered a reflexive practice that could engage them in a more penetrating social analysis of their own historical and class position. It also offered a reflexive practice for me to study these paradoxical assertions about the middle class—both their political paralysis *and* political power, their existential alienations, and the possibilities for revitalizing citizen action and repopulating the naked public square (Borgmann, 1992).

Another epistemological challenge was moving beyond the rationalist assumptions embedded in PAR—that transformation is transforming people's minds, particularly their cognitive and political understandings. Rather, the challenge was to additionally attend to extrarational ways of knowing such as intuitive or embodied knowing that could transform their way of being as well as the way of knowing. The educational process sought to transform normative frameworks by offering new symbolic images, metaphors, or stories and by assessing their power in captivating imagination and fostering energy for change. The educational process also created a space for acknowledging how knowing/acting is shaped by emotions, including the emotions that accompany the death of old understandings, ways of seeing, and patterns of living/working alongside the birth of new understandings, ways of seeing, and patterns for living/working (Macy & Young Brown, 1998; Scott, 1987). Thus, attending to thinking as reason, emotion, and imagination would perhaps make people sensitive to the breaking points of the present system and nourish in them a longing for a new kind of society (Baum in Evans, 1987, p. 246).

Finally, drawing from Deep Ecology, a transformative pedagogy would require the development of an ecological consciousness—an enlargement of the sense of self where the external is understood as part of oneself (O'Sullivan & Taylor, 2004). Such a way of thinking profoundly shifts subjectivities and how people see themselves situated in the world. If human beings recognize themselves as dependent on other life forms and the living system for their very survival, then their identity becomes embedded in the natural world. In essence, there is a change in the ground of ethical thinking, for as Taylor (1989) suggests, there is an essential link between identity, an orientation in moral space, and a sense of place. Further, their identity also becomes embedded in the social as they recognize themselves as dependent on other people for their basic needs and as they ponder the ethical implications of these socioeconomic relations. This view acknowledges adult development both as the paradoxical movement toward autonomy and relatedness. These new ontological and epistemological assumptions have significant implications for sustainability pedagogy.

PAR and Pedagogical Assumptions

As Latapi (1988) suggests, there are three dimensions to PAR: a research dimension, an educational dimension, and an action dimension. Typically, the pedagogic and research functions are fused when a community learns systematic methods for accessing and generating knowledge (de Souza, 1988). To do this, a pedagogy of the question is necessary, where incisive questions generate critique. Peace educator Joanna Macy together with Young Brown (1998) suggest that deeper questions, including questions about our relations to the living whole, can "act as a solvent, loosening up encrusted mental structures, and freeing us to think and see in fresh ways" (p. 47). As an educator-researcher, I provided space for learner-researchers to carry out a social critique of both their personal and social reality and develop explanations through their own research. This space is created when they temporarily distance themselves from daily action and when they are exposed to new ideas from within the sustainability field. Through free and open dialogue, individuals test out their views against the views of other participants and by acting on them within the social conditions they find themselves. Creating a safe place or sanctuary (Lange, in press) allows the "mental comforts and conformities to fall away" and allows participants to stand open to the unknown (Macy & Young Brown, 1998, p. 45). Freire (1997) identifies this process as intervening, not as a betrayal of democracy through imposition, but in the service of democracy that enables learners to become curious, to become unsettled, and to decipher the limit situations of why things are the way they are. This unveils opportunities for hope and counters the typical response of hopelessness and futility when confronted with large social and ecological issues. Briton (1996) calls this a pedagogy of engagement that

> alerts adults to the impersonal forces of modernity that are denying them the right to make responsible decisions and stripping them of their dignity, and engages adults in democratic practices as well as provides them with the communicative and critical competencies to resist the further systematization of culture...and [establish] just and equitable communities of citizens. (p. 115)

In this way, learning is a personal and political encounter, with possibilities for personal and social transformation.

Theory of Social Transformation

Michael Welton (1993) argues that it is the crisis of the lifeworld and ecosystem that has led to the rise of New Social Movements (NSMs) and that adult educators and action researchers need to position themselves as a bridge between educational venues and social movements to retain the radical heritage of adult education. Freire (1970; also Shor & Freire, 1987) discuss the importance of educator-researchers being tactically inside a system, promoting structural change inside a system while drawing support from social movement allies who are strategically outside the system. Thus, the sustainability pedagogy deliberately sought to bring participants into contact with social movement activists.

Further, living systems theory challenges some of the conceptions of social transformation implicit in social movements. The ideas that structures and societies can be reengineered and rationally designed and that bodies and minds can be reconstructed are all part of scientism, that considers matter as passive, mechanistic, and infinitely reconstructible (Ruether, 1992, p. 197). Living systems theory moves from a different premise; that is, we are radically interrelated to all aspects of the living system and it is merely a symptom of our fragmented thinking that perceives humans as separate and able to control external elements. Rather than a reductionist and mechanistic view of the world, living systems theory offers a holistic, organic view where all elements—cells, bodies, social systems, and ecosystems—are all dynamically organized and intricately balanced systems (Macy & Young Brown, 1998). Typically, systems are self-regulating, such as the Earth or our bodies, by constantly receiving feedback and adjusting to maintain "flux-equilibrium" (Macy & Young Brown, 1998, p. 41). When significant stress occurs, these systems either fall apart or adapt by spontaneously reorganizing themselves. When such bifurcation points emerge, both order and disorder are created simultaneously (Capra, 1996). How can this biological understanding of change account for the self-reflexive consciousness of humans and societies with specific legitimations of social power? In other words, how can a power analysis be integrated with a living systems analysis?

Macy and Young Brown (1998), Korten (2006), Hawken (2007), and Edwards (2005) all suggest that the self-organizing of small groups globally—groups that are not centrally organized but diverse, spontaneous, and emergent—constitute a new social change practice. Rather than waiting for key visionary and charismatic leaders, leaders arise organically from within small global communities to create change in their immediate environs. In this way, oppositional resistance founded on social analysis is still an important form of action, but the importance of enacting alternatives as another form of social action ought to be privileged, particularly to drain legitimacy and power from the dominant system.

As a student of Gandhi, Naess (1988) suggests that this can be done by cultivating insight into the oneness of all life and shifting perceptual processes that see separation and difference toward a widening and deepening of the sense of self. Nurturing a consciousness shift on the part of individuals who take on ever expanding circles of identification—from individual ego and self-interest toward an expanded sense of self, particularly an ecological self and a social self—constitutes an alternative cognition. From this way of being one names and acts against a relation of injustice but does so with compassion for those bound by the relation—those implicated in the dominator and dominated aspects of the social relation. In this way, social action has less chance of reproducing the hierarchical and power-over relations that it is challenging. As well, it does not rail at a wall of power indiscriminately, using up all its resources and energy. Like the heron silently watching and waiting in the water, it is an active stillness that is vigilant, aware, and waiting for opportunities. Quickly, suddenly, the heron seizes its fish. So, social change is as much preparation, positioning, and timing as well as finding appropriate opportunities for decisive action (Ming-Dao, 1992, p. 14). Nevertheless, when directly attacked in some way, there has been preparation for an immediate response that stays true to a certain way-of-being rather than reproducing the relation of dominance (Ming-Dao, 1992, p. 33).

Finally, one can borrow the power of existing social forces rather than struggling and grasping for power. Like flying a kite, one can harness the forces that are naturally occurring in a system and using initiative, borrow that energy and redirect it as a way to make change (Ming-Dao, 1992, p. 35). In a fundamental way, this is a reinvention of power as Freire (1970) conceived it, responding to changing contexts. In this way, members of a community become agents of emancipatory change as they expand their sense of self, tell new stories, enact alternative social processes, engage in active stillness together with prepared action, and redirect energies. At a broader level, as all things are connected, the energy that is changed in one social field ripples outward into other social fields and into larger nested systems, creating a social transformation. Intentional engagements in many places can converge and quite suddenly result in the emergence of a new system. Nevertheless, patience is required as chaos can reign for some time before the shape of the transformation reveals itself. As Macy and Young Brown (1998) say, "out of darkness, the new is born" (p. 45).

Engaging in a Critical Living Practice of PAR

From 1998 to 2004, I put this reconceptualization of PAR into practice as part of a course on sustainability, offered through university extension. The course has run five times with a total of fifty-two participants. The participants self-select the course *Transforming Your Working and Living* from a description that indicates that participants will be engaged in understanding the impact of the global economy on their daily life and in transforming their work based on principles that are personally meaningful, contribute to community needs, balance family/individual well-being, and respect the natural world. This creates a community of interest. While I describe the details of the course pedagogy and impact elsewhere (Lange, 2001; 2004; in press), I offer a brief description of key elements here.

The participants begin the PAR process by carrying out a social critique of the issues that brought them into the course. This process assists participants in positioning their experiences within a larger socio-politico-historical context. They also analyze their work genealogy by identifying the key norms that function in workplaces and in the culture at large and produces new understandings about the positive and negative impacts of these norms. While visiting numerous small community businesses, nongovernmental organizations (NGOs), and individuals who are enacting sustainability, they become aware of how sustainability practices can be manifested, the principles and values people hold that inform these practices, and the normalizing elements in dominant practices that eclipse the visibility of alternatives. This exposure often provokes a surfacing of the implicit assumptions they hold, enhancing their capacity for self-reflection and social analysis, and shifts their normative frameworks. They track their consumption, relationship patterns, and work habits as a way to further heighten their consciousness of ingrained cultural norms. By identifying the ethics by which they want to live and resurfacing important values that have been submerged under the deluge of expectations during the life course, they are able to rethink the habits by which they actually live. By exploring numerous ideas from sustainability theory, they are able to envision how these habits might be changed. Tracing how their daily activities, such as drinking coffee or eating a banana or wearing a cotton shirt, have social, economic, and environmental impacts globally informs them about structural power relations of privilege and exploitation. Finally, through several elaborate processes, they create a plan for action that will enact their principles, provide possibilities for personal well-being and social solidarity, and address larger socioenvironmental issues.

In a longitudinal study of the course impact, the participants express that they come to see themselves as acting against the grain or believing in countercultural values. They continually demonstrate a deep awareness of being by describing how they see themselves situated within intimate social relations, global socioeconomic relations, and Earth relations. They describe how they try to foster these entwined relations in just and sustainable ways, whether at the micropolitical level in the home or the macropolitical level through consumption patterns and social movement activities. Having come into contact during the class with individuals involved in various social movements, this often becomes a bridge for their involvement in those social movements or a motivation to become active in other types of civic volunteerism—from political parties to unions to public health—although these involvements ebb and flow over time. They claim that the course provides a map of the people and organizations that can continually nurture their new social analytic capabilities and reinforce new habits. Most importantly, they perceive themselves as part of a larger global movement for sustainability that is working for social and environmental justice. They also assert that it is vital for them to gather regularly for support and accountability as a group and they have chosen on occasion to engage in collective education or action initiatives. Overall, they are continually transforming their way of being in the world, experienced initially through the PAR process, which has persisted over time.

Conclusion

Thus, this critical living practice of PAR weaves the three strands of PAR together with new premises from living systems theory and sustainability theory. As part of sustainability education, traditional origins of PAR offer a strong structural analysis of social and economic systems while sustainability theory offers a strong cultural and philosophical analysis. Living systems theory disrupts modernist, fragmented thought through an ontology of wholeness, connectedness, and paradox. A reconceptualized PAR includes an epistemology that acknowledges deeper levels of change reaching the emotions, body, and spirit and creates an ecological consciousness in addition to transformed cognitive and political understandings. Such an ecological postmodern view of PAR fosters a new ethical and moral sensibility, and imagination for change. Through a pedagogy of questions and engagement, PAR is no longer a tool for education, research, and social action but a way of being that is imprinted throughout the process. The embodied action experienced in a PAR process continues past the educational engagement as participants continue to reappropriate knowledge and evaluate their daily thinking and action to cohere with sustainability values. Rather than seeing themselves as separate and autonomous, they exemplify a sense of deep relatedness to all life forms and a striving to be in just and sustainable relations, despite conflicting tensions and paradoxes. They constantly move between personal-political change and public-political change where they understand their every action and every relationship as part of the larger collective for social change. Thus, PAR can be a moving, elastic, critical, and living process that is not time-specific but is carried within people. Like flowing energy, it circulates outward through their intentions and actions to impact others and is likewise shaped by the actions and intentions of others in this global movement. Like a living form, the end or goal cannot be specified but is indeterminate and uncertain. Nevertheless, these actions can form self-organizing systems that merge to engage in counterhegemonic critique, analysis, and action and then may disperse again before they take another form. This critical living practice of PAR illustrates that a reconceptualized practice of participatory action research is a living web of complexity where person and context are inseparable and new possibilities for knowing, doing, and being are ever emergent, ever creating new possibilities for sustainability.

References

Bohm, D. (1980). *Wholeness and the implicate order.* London: Routledge.
Borgmann, A. (1992). *Crossing the postmodern divide.* Chicago: University of Chicago Press.
Briton, D. (1996). *The modern practice of adult education: A post-modern critique.* Albany: SUNY.
Brown, L. (1981). *Building a sustainable society.* New York: Norton.
Brundtland, G. H. (1987). *The Brundtland report to the world commission on environment and development (our common future).* Oxford: Oxford University Press.
Capra, F. (1996). *The web of life.* New York: Anchor Books.
Carr, W., & Kemmis, S. (1986). *Becoming critical: Education, knowledge and action research.* London: Falmer Press.
Carson, T. (1993). Action research crossing the postmodern divide. Paper presented at the *15th Conference on Curriculum Theory and Classroom Practice,* Dayton, OH.
Carson, T., & Sumara, D. (Eds.). (1997). *Action research as a living practice.* New York: Peter Lang.
de Souza, J. (1988). A perspective on participatory research in Latin America. *Convergence, 21*(2/3), 29–35.
Edwards, A. (2005). *The sustainability revolution.* Gabriola Island, BC: New Society.
Elliott, J. (1991). *Action research for educational change.* Buckingham: Open University Press.
Evans, R. (1987). Education for emancipation. In A. Evans, R. Evans, & W. Kennedy (Eds.), *Pedagogies for the non-poor* (pp. 257–284). Maryknoll, NY: Orbis Books.

Freire, P. (1970). *Pedagogy of the oppressed.* New York: Seabury Press.
———. (1997). *Pedagogy of the heart.* New York: Continuum.
Gadotti, M. (2000). *Pedagogy of the earth and culture of sustainability.* Sao Paulo, Brazil: Instituo Paulo Freire.
Gaventa, J. (1988). Participatory research in North America. *Convergence, 21*(2/3), 19–28.
Hall, B. (1981). Participatory research, popular knowledge and power. *Convergence, 3,* 6–19.
———. (2005). In from the cold? Reflections on participatory research from 1970–2005. *Convergence, 38*(1), 5–24.
Hall, B., & Kassam, Y. (1985). Participatory research. *The International Encyclopedia of Education: Research and Practice, 7,* 3795–3800.
Hawken, P. (2007). *Blessed unrest.* New York: Viking.
Korten, D. (2006). *The great turning: From empire to earth community.* Bloomfield, CT: Kumarian Press & San Francisco, CA: Barrett-Koehler.
Lange, E. A. (2001). *Living transformation: From midlife crisis to restoring ethical space.* Unpublished doctoral dissertation, University of Alberta, Edmonton, Canada.
———. (2004). Transformative and restorative learning: A vital dialectic for sustainable societies. *Adult Education Quarterly, 54*(2), pp. 121–139.
———. (in press). Fostering a learning sanctuary for transformation in adult sustainability education. In J. Mezirow, & E. Taylor (Eds.), *The handbook of transformative learning* (chapter 18). San Francisco: Jossey-Bass.
Latapi, P. (1988). Participatory research: A new research paradigm? *The Alberta Journal of Educational Research, 34*(3), 310–319.
Macy, J., & Young Brown, M. (1998). *Coming back to life.* Gabriola Island, BC: New Society.
Maturana, H., & Varela, F. (1980). *Autopoesis and cognition.* Dordrecht, Holland: D. Reidel.
Merchant, C. (2005). *Radical ecology.* New York: Routledge.
Ming-Dao, D. (1992). *365 tao.* New York: HarperCollins.
Naess, A. (1988). *Self* realization: An ecological approach to being in the world. In J. Seed, J. Macy, P. Fleming, & A. Naess (Eds.), *Thinking Like a Mountain* (pp. 19–30). Gabriola Island, BC: New Society Press.
O'Sullivan, E., & Taylor, M. (2004). *Learning toward an ecological consciousness.* New York: Palgrave Macmillan.
Ruether, R. R. (1992). *Gaia & god: An ecofeminist theology of earth healing.* New York: Harper San Francisco.
Schön, D. (1983). *The reflective practitioner.* New York: Basic Books.
Scott, S. (1997). The grieving soul in the transformation process. In P. Cranton (Ed.), *Transformative learning in action* (pp. 41–50). San Francisco: Jossey-Bass.
Shor, I., & Freire, P. (1987). *A pedagogy for liberation: Dialogues on transforming education.* South Hadley, MA: Bergin & Garvey.
Somekh, B. (2006). *Action research: A methodology for change and development.* Maidenhead, UK: Open University Press.
Spretnak, C. (1999). *The resurgence of the real.* New York: Routledge.
Taylor, C. (1989). *Sources of the self.* Cambridge, MA: Harvard University Press.
Varela, F., Thompson, E., & Rosch, E. (1991). *The embodied mind.* Cambridge: MIT Press.
Welton, M. (1993). Social revolutionary learning: The new social movements as learning sites. *Adult Education Quarterly, 43*(3), pp. 152–164.

PART II

International Contexts: Case Studies of PAR, Education, and Social Change

CHAPTER TEN

Chara chimwe hachitswanyi inda: *Indigenizing Science Education in Zimbabwe*

EDWARD SHIZHA

Introduction

In Zimbabwe, as elsewhere in Africa, local communities had well-developed indigenous knowledge systems for environmental management and coping strategies, making them more resilient to environmental change. This knowledge had, and still has, a high degree of acceptability among the majority of populations in which it has been produced and preserved. These communities can easily identify with this knowledge, and it facilitates their understanding of indigenous scientific processes for environmental management, including disaster prevention, preparedness, response, and mitigation (Kamara, 2007). Community engagement is an ongoing, arduous, and necessary process for developing effective science promotion programs that incorporate indigenous perspectives and epistemologies.

Before the imposition of hegemonic Western "civilization," education in Zimbabwe was purely indigenous and participatory. The community was the school and the teacher. Colonization introduced imperial and capitalist individualistic values that were alien to indigenous communities (Shizha, 2005) and which partially destroyed the foundation of indigenous people's existence. With more emphasis now being placed on acquiring Western values, a new indigenous elite that was created gradually set aside the holistic, lifelong, and utilitarian education system and adopted Eurocentric education as the alternative and more rewarding system since it was associated with capitalist values of economic rewards. However, the valuable nature of African indigenous knowledge remained recognized to the extent that today, the call in most African societies is for a renaissance of indigeneity, albeit in a modified form. This chapter explores the place of community and participatory action research in indigenizing science in

Zimbabwe and how teachers may adopt indigenous participatory methodologies in defining and utilizing science as a community product. The chapter also discusses how indigenous knowledge is practiced outside the formal education system but resisted by professionals in preference to the Eurocentric science.

The Historical and Cultural Location of Participatory Research in Zimbabwe

Zimbabwe has two major indigenous ethnic and sociolinguistic groups, namely, the Shona and Ndebele speakers, constituting more than 70 percent and more than 15 percent of the population, respectively, while several indigenous minority ethnic groups such as the Venda, Tonga, Ndau, Kalanga, and Sotho constitute slightly more than 12 percent of the population (Peresuh & Masuku, 2002). Zimbabweans are essentially a rural people, with nearly three quarters of the population living in communal lands or on commercial farms. Only one quarter lives in towns and very few of these forget their rural roots. They see their true homes as somewhere in the rural area, in the community where they were born. The community forms the basis of relationships between the living, the unborn, and the deceased, and it has a strong importance in indigenous people's spiritual and social lives being a place where a person finds life's meaning. It is a great virtue in Zimbabwean culture to be a community participant and many people view marginalization from the community as the cause of economic, social, cultural, spiritual, and mental poverty. My concern in this chapter is the presentation of the commonality and solidarity that exists as core values in social and economic development among indigenous communities in Zimbabwe. My argument is that indigenous Zimbabweans are not individualistic when it comes to issues of social development. Collaboration and participatory action have always been philosophical and ethical concerns in indigenous communities as depicted by the proverb *Chara chimwe hachitswanyi inda* (An individual cannot succeed by working alone). Despite disruptions, disjuncture, and dislocations resulting from colonization, indigenous people resisted individualism in favor of *kubatana/ukubambana* (togetherness), which is the essence of the philosophy of *unhu/ubuntu*.

The history of indigeneity in Zimbabwe is long, hovering around interrogating metropolitan scholars' pretentions in stifling indigenous epistemologies and methodologies perceived as not independently authoritative, except under metropolitan investigators' guidance and approval (Masolo, 2003). Since colonial invasions, indigenous culture has weathered rapid change (Mkabela, 2005), while indigenous people were redefined based on Anglo-Saxon ethnocentric prejudices. Africans were judged in Eurocentric contexts and not in terms of their own sociocultural realities. Therefore, in order for an individual/community to be admitted into a "civilized" society, that individual/community had to abandon its indigenous practices (Ocholla, 2007), which were central to their socioeconomic development.

Traditional ecological knowledge, which was shared community knowledge, played an integral role in environmental protection and restoration (Shizha, 2007). However, colonial misinterpretations defined it as mythical and mystical despite its applicability to indigenous people's health systems, agricultural production, agroforestry, and biodiversity. In terms of Eurocentric ideology, indigenous ecological knowledge was perceived as "irrational ignorant knowledge" (Shizha, 2005, p. 70). Only positivist explanations from Western philosophy were perceived as the source of scientific knowledge. The belief that Indigenes *need* Occidental scientists was/is patronizing and presumptuous (Coombes, 2007). As a response to neocolonial Eurocentric modernization theory, "critics of African development such as Julius Nyerere of Tanzania argued that any meaningful development could only be achieved if the people's culture and popular knowledges were integrated into the process" (Mulenga, 1999, p. 3). African scholars have become very critical of development programs that marginalize indigenous people and their sciences in favor of hegemonic Western colonizing science that produces "universal histories," define "civilization," and determine "reality" (Semali, 1999).

Colonial definitions of reality produced a mentally colonized generation that does not understand, recognize, and appreciate indigenous knowledge. This is a generation that associates learning with learning from above through indoctrination and banking education rather than in transformative processes. Arguably, this situation has produced an intellectually "colonized" mindset. Moreover, if ever indigenous knowledge was incorporated in people's lives, it was often premised on a distinction that Indigenes are locally embedded while decision makers are exogenous to local circumstance (Coombes, 2007). This entrenches bias in the notion that providers of indigenous knowledge cannot simultaneously perform the role of decision makers and competently advance veritable and authentic reality from such knowledge constructs (Shizha, 2007). Even within today's "inclusionary" science, if ever it is practiced at all, indigenous knowledge has been recast as valued *local* knowledge, something which renders it parochial and, therefore, of limited applicability. Evidence is abundant that indigenous knowledge systems throughout Africa are the bedrock of indigenous communities' survival. Numerous examples (e.g., Kaniki & Mphahlele, 2002) show how indigenous knowledge thrives in traditional community development, farming practices (soil conservation, intercropping, farm rotation), and food technology (fermentation techniques, preservation). These skills, knowledge, and attitudes, were and are still shared, adapted, and refined to sustain communities, and bring development in areas such as healing (e.g., traditional/herbal medicine, physical, and mental fitness), nutrition (e.g., vegetarian cuisine), wealth (e.g., intellectual property, ecotourism), education (e.g., customs, traditions, culture, language), and politics (conflict resolution through *matare* or *indaba* [official meeting forum or chief's court]).

Defining Participatory Action Research in the Indigenous Context

Participatory research involves the people and communities that are researched for the production of knowledge. The community should actively take part in defining their social and cultural realities within their cultural settings. According to Mulenga (1994), it refers to

> an emancipatory approach to knowledge production and utilization. Its main aim is to actively involve the oppressed and disenfranchised people in the collective investigation of reality in order to transform their reality. (p. 11)

Kemmis and McTaggart (1988) add that participatory research is "a collective self-reflective enquiry undertaken by participants in social situations in order to improve...their own social practices" (p. 5). Thus, participatory research simultaneously contributes to basic knowledge in social science and social action in everyday life, and should produce collective, locally controlled knowledge that leads to action on problems directly and immediately affects the people (Mulenga, 1998). I should also add that participatory action involves equalizing power and control in decision-making processes by according voice to all those involved in particular situational activities. Participatory research has a long history and continuity in indigenous communities in Zimbabwe since it has always been part of community life with knowledge production and utilization being always a collaborative activity requiring total involvement of and commitment from community members. The *dare/indaba* or official meeting forum was/is the place where issues were/are debated and shared decisions made for the good of the community. Although there were and are gender power relations in decision making, diverse opinions and voices, irrespective of gender differences, are given audience and space to elicit multiple viewpoints before a consensus is reached. Collective decisions are what rationalize the collective and rural community life in Zimbabwe.

Community participation is equivalent to participation action research, synonymous with indigenous life, which is about what local people know and do and have known and done for generations (Shizha, 2007). The ability to use community knowledge produced from local history and memories provides important skills critical to survival in an African context, thus, what local people know about their environment must be included in the planning and implementation process of "official" knowledge production. If participatory action research is about empowering communities and democratic approaches to knowledge production and utilization, then rural communities in Zimbabwe are deeply involved in it as they deal with their problems utilizing the natural environments in a way that benefits their communities. Thus, the principal issues of inquiry place community life and knowledge acquisition in an Afrocentric dialectic that engages people in critical scientific conversations that solve the health, spiritual, mental, and economic problems that they may encounter. While interacting with the natural environment, they discover knowledge useful to their socioeconomic lives. Essentially, recognition of indigenous knowledge and the role of communities in participatory creation and discoveries should resuscitate African epistemology in the "official" knowledge discourse and enforce a corrective critical theory of African science education and knowledge. It should radicalize elitist approaches to knowledge creation and focus on reconstructive and transformative science that is meaningful and dependable in a democratic society. As Fals-Borda (1988) observes, people's knowledge is a critical recovery of history that has taken years of research to prove its existence and of the enormous value so often ignored and devalued by the so-called expert knowledge. In any case, indigenous knowledge constructions in Zimbabwe have always been based on participatory creativity, and the goal of the participatory process has always been to be as inclusive as possible in collecting views from across the community population.

Unhu/ubuntu and the Concept of Community Participation

Community is an essential component of participatory action in Zimbabwe since it brings people together and imprints individuals with their social and cultural identity. The community is a strong cultural, social, political, and economic institution and the basis of the philosophy of *unhu/ubuntu*, an ethic or humanist philosophy focusing on people's allegiances and relations with each other. Concerning *unhu/ubuntu*, Archbishop Desmond Tutu (1999) elaborates,

> A person with *ubuntu* is open and available to others, affirming of others, does not feel threatened that others are able and good, for he or she has a proper self-assurance that comes from knowing that he or she belongs in a greater whole and is diminished when others are humiliated or diminished, when others are tortured or oppressed.

While some scholars, like Mbigi (1997), describe *ubuntu/unhu* in terms of brotherhood, *unhu/ubuntu* is a multifacetted concept. Samkange and Samkange (1980, pp. 6–7) highlight three maxims of *unhuism* or *ubuntuism* that shape this philosophy: The first maxim asserts that "To be human is to affirm one's humanity by recognizing the humanity of others and, on that basis, establish respectful human relations with them." The second maxim means that "if and when one is faced with a decisive choice between wealth and the preservation of the life of another human being, then one should opt for the preservation of life." The third maxim as a principle deeply embedded in traditional African political philosophy says that "the king owed his status, including all the powers associated with it, to the will of the people under him." *Unhu* embodies all the invaluable virtues that society strives for toward maintaining harmony and the spirit of sharing among its members. *Unhu* banishes individualism and embodies a representative role, in which the individual effectively stands for the people (Mogobe, 2003). The individual identity is replaced with the larger societal identity within the individual. Community is greater than the individual. In the spirit of *ubuntu*, collective unity will see to every person's survival. In turn, every person is expected to be loyal to the collective cause.

The origins of the concept of *unhu/ubuntu* can be found in traditional African culture where brotherhood and collective responsibility are placed above individual initiative and self-sufficiency, and where the group takes care of the needs of individuals. In Zimbabwe, the Shona and Ndebele languages abound with proverbs that express the importance attached to community. For instance, the Shona people have a proverb: *Shiri yakangwara inovaka dendere neminhenga yedzimwe shiri*, which means a clever bird builds its nest with other birds' feathers. And in Ndebele, the expression *isisu somhamb'asingakanani* translates as the stomach of a traveller is small, meaning that the community looks after other people. *Unhu/ubuntu* is an accepted collective way of life. This philosophy of mutuality in social life contrasts with the Western or Eurocentric philosophy of social life that is based on individual initiative and competition (Shizha, 2006). Belief in the interdependence and solidarity of the group means not only that the good is distributed, but also the bad: by harming the other, the self is harmed. In other words, if I debase you, I debase myself (Mogobe, 2003). Looking at the attributes of the collectivism and *unhu/ubuntu* and relating those to the earlier definitions of participatory research, we can see that their orientation toward sharing is typically suited to participatory research involving "authentic participation" and "involvement" (Mulenga, 1998).

Indigenous Knowledge and the Human Experience in Zimbabwe

An important aspect of life in rural Zimbabwe is the extent to which indigenous knowledge is an attribute of a whole range of human experience at the intersection of the complete body of knowledge, know-how, and practices maintained and developed by a collective of people who use the knowledge in their everyday lives. In the traditional worldview, environmental resources (land, water, animals, and plants) are not just production factors with economic significance but they also have their place within the sanctity of nature (Kamara, 2007; Ocholla, 2007). Indigenous knowledge is therefore an essential element of the livelihoods of many local communities. A major challenge that African countries continue to face is how to reconcile indigenous knowledge and Western science without substituting each other, respecting the two sets of values, and building synergy on their respective strengths.

Rural people in Zimbabwe have been using indigenous knowledge from time immemorial. Despite the influence of Western values that sought to colonize their knowledge, the indigenous people in Zimbabwe succeeded in resisting total enculturation into a foreign culture. As reported by Shizha (2007), they have continued to use their traditional ecological knowledge that includes physical sciences and related ethnotechnologies, social sciences, and humanities. As Atte (1989) points out,

> In all those fields, each rural group has developed knowledge encompassing theory, concepts, interrelations, factual data and attributive information of a high degree of accuracy. Such knowledge is so good that such societies have been able to exploit them both for social organizations and productive endeavors to maintain the group. (p. 7)

The process of cognitive mapping, acquiring, coding, and decoding information described by Atte should be studied, recognized, and acknowledged as a possible basis for dialogue and for information exchange with rural farmers, who are often assumed to be ignorant according to top-down technology transfer approaches. Whereas indigenous Zimbabwean ways of knowing have previously been misunderstood and misinterpreted in colonial discourses, indigenous knowledge has their relevance in defining the African personhood (Ngara, 2007).

There are assumptions that indigenous knowledge is not scientific, but what we know about its use among indigenous Zimbabweans is that it has worked for the people and it continues to work for them. Participation in knowledge production has contributed to a peoples' science that serves the people and not the status quo and those who control the validation and legitimation of science. People's science involves what Paolo Freire popularized as alternative approach to learning or investigation (*investigacion tematica*) in which the learner or investigator is committed to cultural action for conscientization (Freire, 1970). Admittedly, cultural action is synonymous with community participation in knowledge production. Rural communities are replete with evidence of scientific principles that are applied in community mobilization and development and recognize science as every part of all people (Mulenga, 1998). For example, the use of *mishonga* (traditional herbal medicines) and traditional healing systems use scientific knowledge that has been constructed and utilized by rural communities for many years. Community-based health practices should not be considered immutable. They are always open to negotiation, creativity, and constant reinvention as new medicinal methods are discovered and the need to conserve some plants becomes desirable.

Conservation and sustainable use of trees and medicinal plants is a community responsibility that is implemented by preventing logging (Ramphele, 2004). These practices, including consultation of *n'anga* (traditional healers) and *vadzimu* (ancestral spirits), are the backbone of indigenous people's health system (Shizha, 2007). Before the introduction of cash crops, such as cotton, sunflower, and tobacco, and the use of chemicals in agriculture, community knowledge about soil type and seed varieties was important to crop production in rural communities. In Zimbabwe, the use of cattle manure, kraal, homestead rotation, and selection of indigenous crops, such as finger millet, have been used to maintain soil fertility. Ignoring the role of indigenous knowledge is tantamount to stripping the indigenous Zimbabweans of their cultural and historical identity.

Science Education, Indigeneity, and Community Experiences

Education science in Zimbabwe is largely Eurocentric, and indigenous science and technology are historically and presently generally unrecognized or extremely discounted. We, as African academics, are stuck in our apologetic frame of mind, demeaning our histories and commemorations, while blindly adopting and advancing the Western theoretical frameworks of science. It is problematic that science education lacks indigenous culture-specific knowledge and ignores the voices of these multicategories of community knowledges. We should feel ashamed of our exclusivist approaches to reading science and to the inappropriateness of our conceptualization of science as stereotypically Western without situating it in particular experiences. This fact is most evident in a simple review of most school, college, and university textbooks; the overwhelming majority does not mention indigeneity, except for an occasional reference to animal life (nonhuman), mineral sources, or plant life (Zulu, 2006). Difference and diversity are at the core of contemporary society and are acceptably pervasive and inevitable attributes of postmodern society. Postmodernism, as a form of inquiry examines knowledge by challenging the monovision and monolithic way of its interpretation. Community-based approaches to science teaching and learning are related to postmodernist influences and treat with respect all versions of cultural and symbolic capital that belong to different social and cultural milieu to prevent the symbolic struggle between antagonistic worldviews (Shizha, 2006). This perspective calls for a critical discourse regarding the utility of indigenous theoretical and philosophical ideas and positioning them at the center of science educational policy formation (Zulu, 2006).

A version of sciences that is reflected in schools and proliferated by policy makers, educational planners, and teachers reinforces the intellectual division of labour that mirrors relations of production. In these relations, knowledge is a social and economic product that disempowers the indigenous students and indigenous communities and serves the interests of the dominant class (Mulenga, 1998). The bigoted modernist mind-set of academic professionals facilitates the domination and exploitation of the "powerless." In Zimbabwe, teachers and educational institutions are reluctant to include communities and their indigenous knowledge in constructing science knowledge. According to Shizha (2005, 2006, 2007), teachers do not see the participation of parents and community leaders in their science curriculum as necessary.

Indigenous participatory action is educational and "educated" professionals should contribute to the development of the community through school programmes that are the "antithesis of the individualistic, elitist, capitalist, competitive culture in which we have been schooled" (Mulenga, 1998, p. 39). As academic professionals we are disconnected from society and continually disregard the role of the community in the decision-making process that affects their children and their welfare. Decision making is important in school programs that involve the participation of local people. Even in instances where we have involved parents, participation has been restricted to discussing tuition fees and provision of infrastructure. Within this limited participatory paradigm, the issue of power relations can be hidden in the discourse of "inclusiveness." Although parents may seem to have some form of voice in the school forum, final decisions and the control of the decision process still remains in the hands of educational professionals. Decisions should not be the total responsibility of government institutions and professional organizations that monopolize power and control, but decision making should be localized and be a down-top or horizontal process.

In Zimbabwean schools, the involvement of parents and other community members in pedagogical practices or participation in implementing science programs is perceived as an invasion of and intrusion on the professionals' authority (Shizha, 2007). Educational institutions and the "intelligentsia gate-keepers," that is, teachers and other academics, continue to practice the politics of difference and the colonial politics of control (Shizha, 2005, 2006) whereby the powerful use their influence to decide on behalf of the poor and powerless. Traditionally and historically, science teaching and knowledge production has been inappropriate, and continues to be inappropriate, because it serves to advance the politics of colonial control. Oppression and external control marginalizes and renders indigenous epistemologies and methodologies systemically and institutionally ineffective. The state and its bureaucracy, through unrealistic or manipulative demands for control, technical competence, and financial accountability, have often played an undermining role to the implementation of indigenous epistemology and ontology (Mazonde & Thomas, 2007). Definitions of science, in Zimbabwe, are dominated by cultural hegemony, cultural biases, and cultural irregularities that demean indigenous people's worldviews. Given the negative impact of inappropriate research with indigenous communities, there is an urgent need for an ethical research approach based on consultation, strong community participation, and methods that acknowledge indigenous ways of knowing (Smith, 1999).

From various fields of study, challenges are now arising as to how science is defined and the nature of science itself as a cultural manifestation. Science is not value-free; it is shaped by our culture. Indeed, as Harding (1986) has argued, those who refuse to question the way science is practiced are avoiding the "scrutiny that science recommends for all other regularities of life" (p. 56). In Zimbabwe, science is viewed as an exclusive and private domain of professionals and the "educated" elite (Shizha, 2007). As a cultural construct, science belongs to a community and not individuals (Semali, 1999). Unfortunately, professionals and educators, in their colonial mentality, are reluctant to engage the community in knowledge production thus marginalizing indigenous ways of knowing. Indigenous knowledge is about cooperation, dialogue, and collaboration in the process of knowledge production. Community-based ownership and control of both scientific knowledge and its application are crucial to participation within local communities (Mazonde & Thomas, 2007). Schools are communities in miniature; hence collaborative creativity should be at the center of the education process. From this perspective knowledge is a process of social interaction that takes place within a framework of participation whereby the learner acquires the necessary skills, tools, beliefs, and values to actively participate in and experience community life. The learning process is constructed around people's everyday life thus contributing to holistic and interconnected experiences. Utilizing people's experiential knowledge gives communities co-ownership of natural resources and knowledge construction (Fischer, Muchapondwa, & Sterner, 2005). Therefore, indigenous epistemology as a representation of the local people, constructed by the people, and controlled by the people themselves, should be reflected in informal and formal learning situations.

Experiential knowledge should, in large measure, be visible in educational settings. School science, in Zimbabwe, negates and invalidates the cultural realities and manifestations, and cultural representations of constructed local realities, which determine people's social being, thinking, behavior, and connectedness (Shizha, 2007). The basic knowledge structures that the average African child brings to school have been collectively constructed and transmitted through a participatory and collectivist model of learning with a community focus (Ngara, 2007). Formal education [schooling] is a contested terrain, which is traversed by competing and contradictory constructions of knowledge. The contestations and contradictions can be overcome if academic institutions in Zimbabwe resist pedagogic hegemony and take cognizance of the importance of the communities in which they are located. Local communities should be utilized as vital resources for knowledge production. Since knowledge is a product of people's sociocultural milieu, communities are the active *cultural fields* for creating scientific knowledge (Shizha, 2006). If the community gets involved and schools can initiate pedagogies that constitute historical representations of difference, these representations would constitute an acknowledgement that knowledge diversity is a feature of the contemporary schooling enterprise.

Community knowledge, which is collectively owned, and community participation are vital to successful educational programs. Participatory methods produce knowledge that is valid and reliable, representing context specific ways of knowing the indigenous world. The most general characteristic of the process of indigenous learning that may itself have an influence on the delivery of education is its contextualization. Indigenous knowledge is typically tied to and incarnated in specific social, cultural, and economic activities within the concerned community, and it is typically acquired by some form of participation in those activities (Semali, 1999). Much of formal and/or organized education in African schools, however, is largely decontexutalized and involves learning things—and learning in ways—that show little relation with the social, cultural, and economic habits of the host community. Centralized, top-down approaches to education and development have left the rural poor increasingly dependent on public provision with no sense of ownership over their own destiny (Chilisa & Preece, 2005). Unquestionably, sometimes there is an inevitable tension between the curriculum planners' values and those of the communities they serve and institutionally pressurized academics requiring a constant trade-off between the ideology of participation and the reality of the educational process (Mulenga, 1999). This problem can be resolved by realizing that indigenous knowledge and school science are socially and culturally constructed and a product of the collective consciousness and actions of people within their communities (Shizha, 2006). As pointed out by Anisur Rahman (1985),

> the distinctive viewpoint of PAR [recognizes that the] domination of masses by elites is rooted not only in the polarization of control over the means of material production but also over the means of knowledge production, including... the social power to determine what is valid or useful knowledge. (p. 119)

The example of Communal Areas Management Programme for Indigenous Resources (CAMPFIRE) discussed in the following text illustrates how integration of indigenous and Western science contributes to better lives for the people.

Community Participation through CAMPFIRE

Zimbabwe's CAMPFIRE is widely regarded as one of Africa's most successful contemporary conservation initiatives involving indigenous people.

It permits the residents of communal lands to share in the benefits generated by wildlife utilization on their lands (Murombedzi, 1999). CAMPFIRE assumes that conservation and development goals can be achieved by creating strong collective tenure over wildlife resources in communal lands. From fauna and flora, communal people engage in extracting medicinal product and herbs that are important to manage their health. Through CAMPFIRE, indigenous people integrate professional advice and their indigenous perspectives to living in harmony with their natural environment but benefit from both at the same time. Indigenous people use their knowledge to manage and control diseases that can affect them and their cattle. The knowledge can be codified and classified according to its usefulness in maintaining ecological balance. For instance, indigenous people know which type of bush and tree is important for treating different types of diseases, which trees cannot be used for firewood and so forth. They also use their indigenous knowledge of the cosmos to plan agriculture activities and tracking animal movements.

Community knowledge and personal observation within the local environment covers pertinent angles often overlooked or neglected by the "objective positivist techniques" that are being taught in schools. Within CAMPFIRE projects indigenous people can predict droughts as well as weather-related diseases by watching the movements of celestial bodies in combination with observing the date of emergence of certain plant species that might affect their crops. Such early warning signals of an approaching environmental disaster are used to determine any preventive measures, prepare for mitigation, and decide on the course of the community in using the natural resources. Similarly, estimates of animal fertility can be drawn from such forecasts with implication on stocking rates and density. This knowledge is little used in schools and less researched so far. The VaTonga and VaKorekore in the Zambezi Valley learn names of the animals and plants, their behavioral patterns, and ecological factors under which they flourish. They keep inventory of species and records of those that disappear. They assign names to new plants and animals. The taxonomy reflects the use of plants for medicinal, social, economic, or cultural usefulness or other determining characteristics, as in the case of poisonous plants. Sometimes biological or ecological features of the species are reflected in the names, such as *muchetura* (for poisonous plant). This taxonomy of important species is then incorporated into everyday knowledge of the community. Knowledge of traditional practices has not yet sufficiently been integrated into the formal educational and health domains in Zimbabwe, a missed opportunity for culturally appropriate programs.

CAMPFIRE has been cast as an antidote to colonial legacy of technocratic and authoritarian development that had undermined people's control over their environment and criminalized their science and use of natural resources (Alexander & McGregor, 2000). From a developmental perspective, CAMPFIRE ameliorates the chronic nutrition status of the indigenous people who had been depending on food aid from donor agencies. The liberatory nature of the program can be discerned from a community leader who was sent to confront local government authorities with their project proposal with these words: *"Tell them that these are our animals and these are our plans. We will not accept any changes imposed by others."* The program has rekindled a proprietorial attitude toward wildlife and natural resources in general. The interaction of indigenous people with wildlife, as promoted by CAMPFIRE, transforms indigenous peoples' lives into a culturally and biologically rich and sustainable future.

Conclusion

Indigenous knowledge is a precious national resource that can facilitate the process of disaster prevention, preparedness, and response in cost-effective, participatory, and sustainable ways. It is an important resource to science education that can be utilized profitably in schools to generate knowledge that is meaningful to students and their communities. Specifically, from time immemorial, natural disaster management in Africa has been deeply rooted in local communities that apply and use indigenous knowledge to master and monitor social development and for survival. Without this knowledge many African communities would simply not be able to survive. Sadly, academic and other professionals in their desire to achieve Eurocentric modernization, bureaucratization, sophisticated economy, and a lifestyle that is closer to that of the Western world have formed an elite group that is not open to accommodating indigenous epistemology and methodologies that favor organized community involvement. Indigenous knowledge provides the basis for problem-solving strategies for local communities and represents an important component of global knowledge on development issues. Incorporating indigenous knowledge into science education and utilizing learning strategies from indigenous perspectives by investigating first what local communities know and have, can improve understanding of local conditions and provide a productive context for activities designed to help communities and making science education an exciting people's science project.

References

Alexander, J., & McGregor, J. A. (2000). Wildlife and politics: CAMPFIRE in Zimbabwe, *Development and Change, 31*, 605–627.

Atte, D. O. (1989). *Indigenous local knowledge as a key to local-level development: Possibilities, constraints, and planning issues in the context of Africa.* Seminar on reviving local self-reliance: Challenges for rural/regional development in Eastern and Southern Africa, United Nations Centre for Regional Development and Centre on Integrated Rural Development for Africa, February 21–24, Tanzania.

Chilisa, B., & Preece, J. (2005). *Research methods for adult educators in Africa*. Cape Town: UNESCO/Pearson.
Coombes, B. (2007). Postcolonial conservation and Kiekie harvests at Morere, New Zealand: Abstracting indigenous knowledge from indigenous polities. *Geographical Research, 45*(2), 186–193.
Fals-Borda, O. (1988). *Knowledge and people's power: Lessons with peasants in Nicaragua, Mexico and Colombia*. New Delhi: Indian Social Institute.
Freire, P. (1970). *Pedagogy of the oppressed*. New York: Continuum.
Fischer, C., Muchapondwa, E., & Sterner, T. (2005). Shall we gather 'round the CAMPFIRE? Zimbabwe's approach to conserving indigenous wildlife. *Resources for the Future, 158*, 12–15.
Harding, S. (1986). *The science question in feminism*. Ithaca, NY: Cornell University Press.
Kamara, J. (2007). Indigenous knowledge in natural disaster reduction in Africa. *The Environmental Times*, A periodic publication by UNEP/GRID-Arendal.
Kaniki, A. M., & Mphahlele, K. M. E. (2002). Indigenous knowledge for the benefit of all: Can knowledge management principles be used effectively? *South African Journal of Libraries and Information Science, 68*(1), 1–14.
Kemmis, S., & McTaggart, R. (1988). *The action research planner* (3rd ed.). Geelong: Deakin University Press.
Masolo, D. A. (2003). Philosophy and indigenous knowledge: An African perspective. *Africa Today, 50*(2), 21–38.
Mazonde, I. N., & Thomas, P. (Eds.). (2007). *Indigenous knowledge systems and intellectual property in the twenty-first century: Perspectives from Southern Africa*. University of Botswana: CODESRIA.
Mbigi, L. (1997). *Ubuntu: The African dream in management*. Pretoria: Knowledge Resources.
Mkabela, Q. (2005). Using the Afrocentric method in researching indigenous African culture. *The Qualitative Report, 10*(1), 178–189.
Mogobe, R. B. (2003). The philosophy of *ubuntu* and *ubuntu* as a philosophy. In P. H. Coetzee, & A. P. J. Roux (Eds.), *The African Philosophy Reader* (2nd ed.) (pp. 230–238). London: Routledge.
Mulenga, D. (1994). *In from the margins: Challenges, possibilities and limitations of participatory research. Summer school in participatory research*. Umtata: Human Sciences Research Council.
———. (1998). In from the margins: Challenges, possibilities and limitations of participatory research. In D. W. Ntiri (Ed.), *Adult education and social change* (pp. 8–49). Detroit: Wayne State University.
———. (1999). Reflections on the practice of participatory research in Africa. *Convergence, 32*(1–4), 33–46.
Murombedzi, J. (1999). Devolution and stewardship in Zimbabwe's CAMPFIRE programme. *Journal of International Development, 11*(2), 287–293.
Ngara, C. (2007). African ways of knowing and pedagogy revisited. *Journal of Contemporary Issues in Education, 2*(2), 7–20.
Ocholla, D. (2007). Marginalized knowledge: An agenda for indigenous knowledge development and integration with other forms of knowledge. *International Review of Information Ethics, 7*, 1–10.
Peresuh, M., & Masuku, J. (2002). The role of the primary language in the bilingual bicultural education in Zimbabwe, *Zambezia, XXIX*(I), 27–37.
Rahman, M. A. (1985). The theory and practice of participatory action research. In O. Fals-Borda (Ed.), *The challenge of social change* (pp. 105–132). London: Sage.
Ramphele, M. (2004). *Women's indigenous knowledge: Building bridges between the traditional and the modern*. Washington: World Bank.
Samkange, S., & Samkange, T. M. (1980) *Hunhuism or ubuntuism: A Zimbabwe indigenous political philosophy*. Salisbury (Harare): Graham.
Semali, L. (1999). Community as a classroom: Dilemmas of valuing African indigenous literacy in education. *International Review of Education, 45*(3–4), 305–319.
Shizha, E. (2005). Reclaiming our memories: The education dilemma in postcolonial African school curricula. In A. Abdi, & A. Cleghorn (Eds.), *Issues in African education: Sociological perspectives* (pp. 65–83). New York: Palgrave Macmillan.
———. (2006). Legitimizing indigenous knowledge in Zimbabwe: A theoretical analysis of postcolonial school knowledge and its colonial legacy. *Journal of Contemporary Issues in Education, 1(1)*, 20–35.

———. (2007), Critical analysis of problems encountered in incorporating indigenous knowledge in science teaching by primary school teachers in Zimbabwe. *Alberta Journal of Educational Research, 53*(3), 302–319.
Smith, L. T. (1999). *Decolonizing methodologies: Research and indigenous peoples.* London: Zed.
Tutu, D. *(1999). No future without forgiveness.* New York: Image.
Zulu, I. M. (2006). Critical indigenous African education and knowledge. *Journal of Pan African Studies, 1*(3), 32–49.

CHAPTER ELEVEN

Research and Agency: The Case of Rural Women and Land Tenure in Tanzania

CHRISTINE HELLEN MHINA

Introduction

Women in Tanzania constitute the majority of smallholder farmers but due to the structure of the rural society and its related customary law, the majority of women have inferior and insecure land rights relative to men (United Republic of Tanzania [URT], 1995). Adult women with children but without a spouse, particularly in rural areas, face the greatest difficulty in handling the land problem. An increasing incidence of unwed mothers, widows, and divorcees makes this problem even more widespread that it has been in the past.

The problem of women's rights to land in Tanzania has been so difficult to resolve (Mhina, 2005; Toulmin & Quan, 2000). Despite legislative changes and considerable efforts of the national government, little has changed for women at the village level. In addition, the challenge that rural women face in confronting the problematic land issue is far more complex than the literature has usually implied. One of the main reasons that women in the villages have difficulty with this issue is that they do not fully understand how the situation is maintained or how to change it. In addition, no attention has been given to the involvement of women at grassroots levels in their struggles to ameliorate the significant economic handicap for women farmers (Mhina, 2005). In other words, those who are most affected have largely been left out and have been denied the opportunity to develop a sense of agency within their communities.

Tanzania's land policy has been in a state of crisis for the past three decades. In the early 1990s, Tanzania as a nation actively engaged in debates about reforming the land tenure structure. However, due to oversimplification of the matter by policy makers, the issue of gender and land rights has received little attention in policy formulation (Manji, 1998).

The neglect of women's land-related concerns by governmental institutions mirrors a gap within academic scholarship, where the relationship between women and their land rights is virtually unattended and little theorized. Although research concerning the impact of customary patterns on women farmers in Tanzania has expanded, very little attention has been given to the involvement of women at grassroots levels in their struggles to ameliorate this handicap for women farmers.

The focus of this chapter is not on land problems per se but on the learning process that rural women went through while engaging in dialogue and actions directed toward seeking local solutions to the land problem through the use of participatory action research (PAR). PAR in this case study was a means of accessing untapped potential existing at grass roots level, which is central to producing needed social changes in rural Tanzanian society. I argue that when provided with the opportunity to share and collectively reflect on their experiences, people at the grassroots level directly affected by problems (in this case women with marginal land rights) are capable of understanding and acting on the issues, equally if not more effectively than other more distant players.

Is Women's Participation a Valid Concept?

In early 1990s the concept of participation was emphasized in development literature. Seemingly it was more so in thinking than in practice (Chambers, 1983; Nyerere, 1979; Oakley, 1991; Rahnema, 1990). For instance, Oakley (1991) argued that people in communities have been dominated by and have been dependent on local elite groups for generations and this dependence is the most powerful barrier to people's participation in development activities. Rahnema (1990) clarifies this dependent relationship, which he thinks is created by the assumed superior position of professionals in society. He argues that "we as professionals view ourselves as interveners and think we have the answers to the problems of the intervened" (p. 205). Kapoor (2005) dwells on concerns that pertain to the use of nongovernmental organization (NGO) interventions to stifle people's protest and activism in relation to processes of development-marginalization. It is the NGO and allied players like lawyers, development practitioners, and policy makers taking the lead in making decisions on how to address the persistent women's land issue. They do this without considering the fact that by not letting people face up to their own problems, they cripple rural women's capacity to make their own decisions on issues that concern them (Horton & Freire, 1990). Park (1993) shares a similar view:

> By not allowing people to make their own decisions, the traditional sources of human strength and capacity are taken away and replaced with incapacitation and helplessness. (p. 16)

As a result they succumb to dependency, letting experts from the government, NGOs, or other agencies decide and act for them (Horton & Freire, 1990). Through this process, they become nonparticipants in decisions that affect their lives. As Kapoor (2005) rightly asks, isn't this jeopardizing prospects for promoting people's agency and political assertion?

The advocacy for people's agency and participation in development started in late 1960s. For instance, Nyerere (1979) emphasized development approaches that enhance people's strong sense of agency characterized by a belief in their ability to master circumstances. In a similar vein, Schumacher (1973) argued that without efficient utilization of people's knowledge in terms of their education, organization, and discipline, these resources would remain latent, untapped potential. It means that without involving people in their own development, their skills and talents will be wasted. This line of argument supports the assertion that rural people are not ignorant, idle, or apathetic, as they are often made out to be; on the contrary, they are resourceful, knowledgeable, and hard working (Chambers, 1983; Freire, 1970; Mhina, 2005; Nyerere, 1979; Oakley, 1991), when given the opportunity to use their skills.

Advocating for Women's Local Knowledge

In his theory of structuration, Giddens (1984) talks about knowledge that human actors possess. He believes that by virtue of active participation every member of society gets to know a great deal about the workings of that society. He emphasizes that all competent members of society are vastly skilled in the practical accomplishment of social activities. Giddens (1984) writes, "The knowledge of social conventions, of oneself and of other human beings, presumed in being able to go on in the diversity of contexts of social life is detailed and dazzling" (p. 26). He further argues that most of the rules implicated in the production of social practices are not in explicitly codified form. Instead, as Polanyi (1962) argues, social actors grasp those rules tacitly through nonexplicit process of knowing. Thus, the core of social actor's knowledgeability is their awareness of social rules, which is expressed in practical consciousness.

The nonexplicit process by which we know and do things accounts for the possession by humanity of an immense mental domain of knowledge, particularly of the many different arts that people know how to use, enjoy, or live by, without knowing their contents (Polanyi, 1962). In his notion of tacit knowing, Polanyi showed that people's implicit knowledge could be revealed either by performance of intelligent actions. He argues that there are some experiences that people grasp but cannot speak of or describe, nor can they even entertain in conscious thought the knowledge that their actions reveal. Thus, human beings as social actors are highly learned in respect of knowledge, which they possess, apply, and comply with (Giddens, 1984). Without being aware of or able to express the knowledge that is tacitly embedded in tradition and culture, people use it as an unarticulated background against whatever they currently attend.

Despite the above assertion on social actors being knowledgeable, we professionals are still grappling with building faith in ordinary people. For instance, feminist authors writing on issues of women and land rights argue against the possibility of the powerless to negotiate for their rights and point to the fact that "not everyone is able to be an interlocutor and may lose in such negotiations and conversations" (Peters, 2004, p. 279). Others opine that "women have too little political voice at all decision-making levels that are implied by the land question: in local-level management systems, within the formal law and also within the government and civil society itself" (Whitehead & Tsikata, 2003, p. 104). Such writings obstruct an understanding of the true potential of African rural women and underestimate them.

Chambers (1995) reminds us that it is the perspective of those who have been left out of the development process that matters. If we were to see rural women as knowledgeable people, it would have been easier for us "professionals" to encourage them to take initiative in addressing their common concerns (Mhina, 2005). It is thus hard not to agree with the authors of World Commission on Culture and Development Report in their call to adopt new ways of thinking instead of depending on conventional answers. Here I quote from UNESCO (1988):

> The world as we know it, all the relationships, we took as given, are undergoing, profound rethinking and reconstruction. Imagination, innovation, vision and creativity are required...It means an open mind, and open heart and a readiness to seek fresh definitions, reconcile old opposites, and help draw new mental maps. (p. 12)

This was a call to see things the other way around, to develop new ways of thinking, acting, and organizing, and working directly with the poor to encourage and enable them to learn to become active participants again (Mhina, 2005). In similar vein, Chambers (1994) urges professionals to embrace the new paradigm, as he emphasizes "to see things the other way round, to soften and flatten hierarchy, to adopt downward accountability, to change behavior, attitudes and beliefs, and to identify and implement a new agenda" (p. 14). As Freire (1970) reminds us, women's subordination and marginalization is not a given destiny. We professionals are challenged to commit ourselves to catalyze rural women's own ability to assert their self-interests; otherwise their access to resources will remain marginalized. I argue that participatory action research (PAR) has a significant role in facilitating this process.

Participatory Action Research

Participant Selection and Their Identification of the Problem

This PAR case study was undertaken at Maruku village in the Bukoba District of the Kagera Region in Tanzania. My desire was to work with the most sharply affected by inability to access and control agricultural land to take part in a process of determining how to change their problematic situation. The process of identifying participants and enticing their willingness to participate in learning circles entailed building a trusting and noncontrolling relationships with and among participants. At the initial level of gaining entry into the village, I identified myself with female agricultural extension officers located in the area and established friendship with them. We were visiting one another frequently; their neighbors and relatives became my acquaintances as I continued to expand my circle of associates. I attended different cultural functions, including weddings and funerals. Soon I was treated as one of the members of their community and the leadership in the village was very supportive in the process of connecting with women.

Selection of potential participants was accomplished through one-to-one interviews, which was done in a storytelling method. I let the women who were willing to participate in the PAR process describe what they perceived to be their problem. They identified themselves as mostly affected not only by the lack of access to and control of the agricultural land but also by the complex process of seeking solutions to the problem. The group of participants comprised of ten women: six widows, two married women, one divorcee, and one single unmarried woman. The age difference ranged from thirty-three to sixty-five years and schooling ranged from no schooling at all to fourteen years of formal schooling. Despite these demographic differences, these participants were not distinguished by any major differences that signify social or political differentials. Such a commonality, allowed the environment of mutuality and comfort, which encouraged them to share their personal experiences.

Collective Learning through Searching and Reflection

Collective Search for Data

Data was collected in two phases. In the first phase, participants collected data through (1) sharing their experiences, (2) listening to the trigger story (see the following text), (3) consulting community elders and village (government) authorities, and (4) consulting the Tanzania land policy and law as related to women's rights to land. The second phase was basically my observation and documenting of the learning process the participants were involved in.

The working procedure for data collection was interactive group dialogue; however, before the dialogue sessions I interviewed potential participants individually to identify those women who would be involved in the sessions. I let women describe what they perceived to be a problem in terms of their rights to land. Based on their own description of the problem and willingness to participate in a group dialogue, I selected ten women participants for the group dialogue. Thus, the same members of the dialogue group met ten times and each participant had a chance to share her experience with the group. Three participants narrated their stories each day during the first three days while others were listening and asking questions. The tenth participant, whose story I considered to be a trigger story, shared her experience on the fourth day.

Drawing from Freire (1970), I came up with the idea of having a trigger story to stimulate group members' discussion. A trigger refers to a code or a concrete representation of an identified community problem in the form of a photograph, video, or a story. In this way, I used the tenth participant's experience as a trigger story to stimulate further participants' discussion of the land problem they encountered. I selected this story because it was the most salient among the stories, and it provoked an emotional response from the group members and inspired women to talk freely about their feelings. In the following text, I present a summary of the trigger story as the participant narrated it to me during the individual interview before the group dialogue sessions.

> Rosanna, sixty-three years old, was a sonless widow who had been married to Patrick for more than forty-five years. Then, in 1995, the husband got sick and was bedridden for five years, during which time Rosanna took intensive care of her sick husband. Unfortunately, during his sickness, their daughter died and was survived by a five-year-old girl who was to be raised by Rosanna. Two years later, Rosanna's husband died too.
>
> Patrick's death brought changes in Rosanna's life. Ordinarily, when a person dies within Patrick's tribal group, a special meeting (*matanga*) is held four days after the funeral. During this meeting, the clan council disposes of the properties, claims, and debts of the deceased and also decides on the fate of the widow as it is guided by the rules and customs of the tribe. It was during this meeting that the clan council took a decision to evict Rosanna from the place that had been her home for more than forty-five years. The fact that she lived there throughout her adult lifetime and worked on her husband's farmland (*kibanja*) during her married lifetime and took care of her sick husband without any help from the relatives was entirely disregarded.
>
> The news of her eviction came as a complete shock to her. She was appalled and did not know what to do. Going back to Ngara, her natal home was not an option because she had left her parental home when she was only fifteen years old, moved to Bukoba, and got married to Patrick. She had never gone back to Ngara since, even for a visit. She

feared that she might not be received in her natal village. Although Rosanna continued staying in her husband's *kibanja,* she led a miserable life and was extremely insecure. Rosanna lived in an old dilapidated house. The walls on one side of the house had already collapsed and everything was falling apart, totally unsuitable for human habitation. Rosanna was forbidden to harvest bananas from the *kibanja* even for her young granddaughter and her own consumption. She had no support whatsoever, neither from individuals nor from the village government. She struggled in solitary to cope with ever intensifying poverty.
—Collective Data Analysis through Reflection

Data analysis was also performed in two phases. First, it was participants' own reflection (a thinking through) process about their situation and strategies for changing their situation. Participants reflected on their shared information, the trigger story, and also laws and policies. As they were solving the jigsaw puzzle, they made several decisions, including the decision to consult others (elders and village authorities), for clarification of what was not clear to them. The consultations led the participants to further reflections and researching for more information. It was a cyclical process of searching and researching. This was the beginning of collective strength.

The second phase of analysis was my own assessment of the changing nature of participants' ideas, interactions, and actions as they worked together throughout the sessions. I observed (1) the changing nature of their connectedness, (2) the changing nature of their knowledge, and (3) the changing nature of their actions. For instance, during the analysis of the trigger story, it was interesting to observe the journey they went through cognitively from the moment Rosanna introduced her story, through the discussion to the point they implemented their social action.

Mutual learning

In this PAR process, the participants and I engaged in mutual learning. While participants shared their stories, it was important for me to respect their thinking and intuition and so I attentively listened to them and I was open to learn from them. I viewed my role as one of creating a space where people could speak openly on their own behalf (Lather, 1991). I also probed to provide a context for participants' critical reflection and to benefit from the breadth of their heterogeneous experiences. Obviously, participants had insider knowledge from their experiences, while I shared information from my professional experience and theoretical perspectives.

Women's Agency, Activism, and Inarticulate Intelligence

"We Are Powerless"

Participants broke their silence by coming together to talk about their experiences through dialogue and after a course of ten dialogue sessions, their social understanding of their situation had changed. They used their personal experiences as a starting point to critically reflect and raise questions about prevalent forms of injustices that appeared to be deeply entrenched in the system. All participants had attempted to solve the land disputes they encountered without success. At the time as I was conducting this research two participants were still pursuing their land disputes at the District Court in Bukoba; four women had despaired after unsuccessful struggles of seeking solutions, a phenomenon that led them to join other women (nonparticipants) who had opted to keep silent. At the initial stage of sharing their experiences, participants' stories converged and they began to recognize their commonalities. The following text is a summary of issues that defined the foundation of their common struggle.

Women's stories depict a condition of dependency. Throughout their lives, these women lived as dependants of their male relatives. They were insecure and lived with constant fear of becoming homeless. Since a married woman in Bahaya tribe ekes out her livelihood from her husband's land, her security could be severely threatened by divorce. Being divorced means losing not only a husband and children, but also access to a land resource. Seemingly, one would prefer to stay in a marriage irrespective of the misery that this institution might be loaded with. One participant commented, "we are born to suffer, do we have any choice, this is our destiny." In some cases widows are forced to return to their own clan-land, which is their parents' home. But returnees are regarded as trespassers, particularly when the land resource is kept under the custody of the brother or any designated male trustee. A widow who was evicted from her parents' home commented, "When my parents sent me off for marriage, I did not realize I was considered as gone and nonexisting."

Women who were pursuing their court cases were dealing with the frustrations of making several return trips from the village to the town (approximately twenty-four kilometers) on foot, basically to sit outside the courtyard waiting for a court clerk to reschedule dates for unforthcoming court hearing. They were totally uninformed of how the foreign court system operates. At times they had to fight various forms of intimidation, including direct physical violence and threats of being killed by their opponents. One participant who has been fighting for years to acquire effective control of her father's *kibanja* was brutally assaulted by her opponent (a male kin) on her way home from the court and ended up being reprimanded for wrong accusation. Seemingly these women were fighting for justice indefinitely with no hope of success.

Can We Do Something?

Although at the beginning the women's stories depicted their powerlessness and inability to pursue their interests, their participation in the rigorous PAR process proved the opposite, giving strong evidence that people are made to believe they are incompetent, deficient, incapable, and irrelevant (Kapoor, 2005). On the contrary, participants adopted a critical view of their situation, that of people who question, who doubt, who investigate, and who want to illuminate their lives. Participants looked at their realities in a new way and participated in knowing their reality on a higher and more critical level. They developed a better understanding of the unjust structures that underpinned their marginalization and subordination. They took the initiative to intervene in the state of affairs to transform the social situation in which they found themselves. Giddens (1979) refers to this as agency, which only exists when an agent has the capability of intervening or refraining from intervening. Participants were exercising their agency together as a group, at the same time displaying various abilities as described in the following text, as they continued with their reflections.

Critical and Analytical Skills

According to Kemmis and McTaggart (2000), participatory action research emerges in situations where people want to make changes thoughtfully; that is, they think realistically about where they are now, how things came to be that way, and how, in practice, things might be changed. Participants were more than able to analyze their situation. During the discussions of each other's story and particularly the trigger story, the women analyzed clearly how each case was handled by family members, the village authorities, the courts, or other stakeholders and what could have been done differently. They looked at the land problem from both the perspective of individuals and also the broader perspective of wanting to know how each case was connected to wider social and historical conditions (Kemmis & McTaggart, 2000). As Freire (1970) argued, "When men lack a critical understanding of their reality, apprehending it in fragments, which they do not perceive as interacting constituent elements of the whole, they cannot truly know that reality" (p. 95).

Often, participants asked how the system operated both at the village level and at the level of wider society. Participants had the opportunity to reflect on laws and policies as they appeared in text and to see how they were implemented. As the dialogue progressed, participants began to see the larger picture of the structural contradictions that caused their social, economic, and political miseries. They came to realize that this was not an individual problem, but a social problem faced by many women in their home districts. They also realized that for most women, effective rights in land remain elusive, as their male heirs were unlikely to relinquish their privileges of inheriting the land resource. The participants came to agree that women in general are not informed of the rights that contemporary laws have promised them.

The discussion of the trigger story centered on the question of why Rosanna should be evicted. It was beyond the women's imagination that Rosanna could be evicted from her home and be forbidden to harvest from her own farm. To be forbidden not to touch anything from that farm was a serious act of severance. They felt that Rosanna was betrayed and her eviction and subjection to misery was unjustified. Throughout their reflective journey, their main concerns were: Why was Rosanna evicted? What is really happening? Why do not things work out as anticipated? Can we do something about it? Are we going to do something about it? What are we going to do about it? It was if the culture of silence was suddenly shattered. They discovered that they could speak not only about their problems but also understand that things have to be made different by themselves and others. Participants, through their powerful arguments and their critical reasoning were able to influence each other's passion to a determination and hence to implement collaborative ventures.

We Can Make the Change Happen!

Kemmis and McTaggart (2000) recognize the primacy of practical reason—the necessity of deciding what to do when confronted with uncertain practical questions. Similarly, participants in this case study were dissatisfied with the way things were, and they continuously demonstrated progressive attempts to make a difference. Through practical reasoning they made various decisions as they sought solutions for their problem. For instance, during the reflection of the trigger story, participants were affected by the way the clan council handled Rosanna's case. They came to a point of wanting to counteract all the atrocities she suffered. Then based on a shared and passionate understanding of their situation participants stood up together and took control over what they needed to work with, and in their own way. They agreed to rebuild Rosanna's house and to confront Rosanna's opponent collectively to hear the other side of the story. Given their marginalized position in the society, they figured out that they needed to work collaboratively with leaders, who were mostly men. They asked village authorities to revoke the decision of the clan council and to review Rosanna's case from the beginning. They also suggested having few participants participate in the review committee. Indeed, participants outwardly projected their enthusiasm and at this point their determination was obvious.

Toward the end of the dialogue sessions, participants established a solidarity group named *Tweyambe,* meaning let us work together to safeguard women's rights. *Tweyambe* was a tool and means for women to begin the work of settling disputes among themselves rather than relying on traditional rural power structures. Through *Tweyambe* women have the opportunity to engage in the analytical process by which structural impediments come more clearly into view. *Tweyambe* was also a tool for women to exercise their assertion for their own independent voice. With *Tweyambe,* women lost fear and they gained self-esteem. All these decisions and actions were collaboratively organized because participants did not work as individuals; rather they worked as a team. They were sympathetic as they sought connections with each other to build a sense of community. Solidarity was as a source of strength for their struggle, and it was the compassionate relationship that strengthened their commitment to a shared struggle.

Borrowing from Kabeer (1999) who uses the term resources to encompass various human and social resources, I consider the abilities (analytical and critical skills, creativity, solidarity, and networking) participants displayed as resources that enabled the group members to move from one level of understanding to another and to act on their situation. Their decisions on how to act in a right way in a given circumstance were based on practical knowledge of the social rules that they grasped tacitly. It was their collective mental power and their organizational capacities that they possessed, although they did not know that they had such resources. It could be said that they made explicit the form of intelligence that was inarticulate (Polanyi, 1962).

It suffices here to say that women had gone through a process of awakening. Participants were initially timid but later became stronger. The change in women's understanding was a movement from a lower level of understanding to a higher level. This change in understanding can be described under three dimensions. The first dimension denotes a change in knowledge from individual experiential knowledge of the problem to recognition of the complexity of the entire land issue. This is what Freire (1970) calls a verbal unveiling of their reality. They recognized and denounced their marginality by thoroughly reevaluating and reinterpreting their own experiences.

The second dimension of the movement denotes a change in terms of group members working individually to a collective action. At the beginning, group members were trapped in their powerlessness and isolation and later they were able to break this powerlessness and fight for a common goal to change their situation. Participants realized that it works better for them to work collectively, with social solidarity and interactive knowledge as fundamental elements for their empowerment.

The third dimension denotes a movement from inactivity to acting. Women's critical discourse was useful in organizing what action to be taken. The popular saying that activity overcomes passivity is relevant here. Participants explored ideas and instead of waiting to be told what to do, they established their own association, *Tweyambe*. They also identified allies, particularly the elders and village authorities who seemed to share beliefs, attitudes, and values that are consistent with building social justice. By exchanging views with these people, group members came up with concrete decisions about what to do and how to deal with their problems. Thus, by taking part in the PAR process, participants revealed their agency, activism and inarticulate intelligence, which had been encapsulated in their condition of uneasiness and misery.

Conclusions

The PAR process in this case study traced the developing understanding of ten rural women as a group during the course of ten dialogue sessions they had with me, the researcher. These women broke their silence by coming together to talk about their experiences through PAR dialogue. Using their knowledge and skills, they progressively saw themselves as being capable of questioning, acting, and changing what was previously forced on them. While the PAR process enabled these women to move from one level of understanding to another and to act on their situation, it provided an opportunity for me to observe the land problem from their point of view while at the same time helping them see themselves as knowledgeable people (Maguire, 1987). This was a significant learning process central to producing needed social changes in rural Tanzanian society.

This case study provides convincing evidence of women's agency and activism. They were potentially capable of dealing with the problem of marginal rights to land both intellectually and sociopolitically. My role as a PAR researcher was to inspire women to reflect on the marginal land rights that affected their lives and to act upon their understanding and decisions to change their situation. Methodologically, it could be said that this entire research process has responded to the experiences and needs of disempowered rural women. The knowledge generated was not given to them and the change in their understanding of their situation was not done for them. Instead it was a consequence of their own effort to engage in a collective struggle to change their situation. Since the knowledge generated by group members is based on people's customs and traditions (Fals-Borda & Rahman, 1991), it is more appropriate and useful in their situation.

The results of this research are useful for community development approaches. Since rural women farmers have a crucial role as direct subsistence producers, reformers of land tenure systems that disadvantage women need to draw from the practical knowledge of those affected by the problem in question. Reformers of land tenure systems and other community development workers need to assume the catalytic role to stimulate dialogue among the people whom the reform is intended to benefit and tap into their practical knowledge of the problem at hand. As Lather (1991) has emphasized, our task as PAR researchers is to take away the barriers that prevent people from speaking for themselves and for us to act as creators of space where people speak openly on their own behalf. Likewise, the results are useful in academia particularly now that there is emphasis on incorporating diverse voices for a successful mutual learning.

Finally, the case study supplies strong support for policy development, which recognizes, honors, and develops local capacities as the best means of addressing pressing social and environmental issues. As Chambers (1998) argued, "What cannot now be repeated is any assertion that the poor are incapable of their own analysis, or any assertion that the powerful lack the approaches and methods to enable them to undertake that analysis... There are now fewer excuses than ever before for ignoring the needs and priorities of the poor" (p. 200).

References

Chambers, R. (1983). *Rural development: Putting the last first*. London: Longman.

———. (1994). Foreword. In I. Scoones, & J. Thompson (Eds.), *Beyond farmer first*. London: Intermediate Technology.

———. (1995). The professionals and the powerless: Whose reality counts? *Choices: The Human Development Magazine, 4*(1). New York: United Nations Development Program.

———. (1998). Afterword. In J. Holland, & J. Blackburn (Eds.), *Whose voice? Participatory research and policy change*. London: Intermediate Technology.

Fals-Borda, O., & Rahman, M. (Eds.). (1991). *Action and knowledge: Breaking the monopoly with participatory action-research*. New York: Apex Press.

Freire, P. (1970). *Pedagogy of the oppressed*. New York: Continuum.

Giddens, A. (1979). *Central problems in social theory: Action, structure and contradiction in social analysis*. Berkeley: University of California Press.

———. (1984). *The constitution of society: Outline of the theory of structuration*. Cambridge, MA: Polity.

Horton, M., & Freire, P. (1990). We make the road by walking: Conversation on education and social change. In B. Bell, J. Gaventa, & J. Peters (Eds.), *We make the road by walking*. Philadelphia: Temple University Press.

Kabeer, N. (1999). *The conditions and consequences of choice: Reflections on the measurement of women's empowerment*. Geneva: United Nations Research Institute for Social Development.

Kapoor, D. (2005). NGO partnerships and the taming of the grassroots in rural India. *Development in Practice, 15*(2), 210–215

Kemmis, S., & McTaggart, R. (2000). Participatory action research. In N. K. Denzin, & Y. S. Lincoln (Eds.), *The handbook of qualitative research* (2nd ed., pp. 567–605). Thousand Oaks, CA: Sage.

Lather, P. (1991). *Getting smart*. New York: Routledge.

Maguire, P. (1987). *Doing Participatory research—a feminist approach*. Amherst: University of Massachusetts.

Manji, A. (1998). Gender and the politics of the land reform process in Tanzania. *Journal of modern studies, 96*(4), 645–667.

Mhina, C. H. (2005). *Social learning for women's empowerment in rural Tanzania*. Unpublished dissertation, University of Alberta, Edmonton, Canada.

Nyerere J. K. (1979). Adult education and development. In H. Hinzen, & V. H. Hundsdorfer (Eds.), *The Tanzanian experience: Education for liberation and development*. UNESCO Institute for Education, Hamburg: Evans Brothers London.

Oakley, P. (1991). *Projects with people: The practice of participation in rural development*. Geneva: International Labor Office.

Park, P. (1993). What is participatory research? A theoretical and methodological perspective. In P. Park, M. Brydon-Miller, B. Hall, & T. Jackson (1993), *Voice of change*. Westport, CT: Bergin & Garvey.

Peters, P. E. (2004). Inequality and social conflict over land in Africa. *Journal of Agrarian Change, 4*(3), 269–314.

Polanyi, M. (1962). *Personal knowledge: Towards a post-critical philosophy*. Chicago: University of Chicago Press.

Rahnema, M. (1990). Participatory action research: The "last temptation of saint" development. *Alternatives, 15*(2), 199–226.

Schumacher, E. E. (1973). *Small is beautiful*. London: Abacus.

Toulmin, C., & Quan, J. (Eds.). (2000). *Evolving land rights, policy and tenure in Africa*. London: DFID/IIED/NRI.

UNESCO. (1988). *Our creative diversity: Report of the world commission on culture and development*. Geneva: Oxford and India Book House.
United Republic of Tanzania [URT]. (1995). *Ministry of lands, housing and urban development: The national land policy*. Dar-Es-Salaam, Tanzania: Printpak (Tanzania).
Whitehead, A, & Tsikata, D. (2003). Policy discourses on women's land rights in Sub-Saharan Africa: The implications of the re-turn to the customary. In S. Razavi (Ed.), *Agrarian change, gender and land rights* (pp. 67–112). Oxford: Blackwell.

CHAPTER TWELVE

NGO-Community Partnerships, PAR, and Learning in Mining Struggles in Ghana

VALERIE KWAI PUN

Ghana's current gold rush has exploded at an extraordinary rate and magnitude. Economic restructuring has lubricated an unprecedented influx of transnational capital into the country and fertilized the proliferation of corporate-led surface mining projects. Communities affected by multinational mining typify the underdevelopment paradox common to most extractive-based economies and live in aching poverty despite their enormous mineral wealth. Mining has displaced tens of thousands of local inhabitants and caused substantial environmental damage to surrounding lands, forests, and water bodies. Growing frustration and disappointment with mining policies and practices have galvanized a collective response from communities and their activist supporters, who demand fair compensation and people-centred development initiatives to improve the quality of life in these areas, including demands for farmable lands, hospitals, schools, and potable water.

As these communities engage in a process of struggle, they embark on a complex and contradictory journey toward social change that is fundamentally educative. Movement participants engage in a process of learning that challenges and significantly alters their understanding of the world, creating the potential for social change (Foley, 1999). The centrality of learning in shaping and directing Southern and recently independent subaltern countries' struggles and social movements (Kapoor, 2007; Kapoor, 2008; Kwai Pun, 2008) underscores the significance of repoliticization of increasingly silenced marginalized social groups through processes of Freirian (1970) popular education (PE).[1] Participatory Action Research (PAR) is central to this learning process as a framework of inquiry that attempts to demystify, reclaim, and decolonize research as an exercise of people's investigation and learning to address the social structures of domination (Fals-Borda, 1988; Rahnema, 1992; Smith, 1999). PAR acknowledges the links between power and knowledge and understands pedagogical processes and research as being fundamentally political (Freire, 1970). In addition, PAR seeks to disassemble traditional mis/conceptions about research (as an absolutist, omnipotent, and ahistorical practice of privilege and expertise) to unearth a significant and ever-present process of people's engagement with knowledge. This process is accessible, arises from conditions of oppression, is ongoing and persistent, and is a form of knowledge generation/utilization that is mostly disregarded in the dominant discourse on research and knowledge production in academia.

Relying on research that began in 2005 (Kwai Pun, 2008) and my observations and experiences as a movement supporter, this chapter will provide a brief description of Ghana's mining sector. Then, it will outline some of the socioecological impacts of mining before examining the role of PAR and learning in community responses to large-scale mining, as well as its role within NGO-community partnerships in the struggle to advocate for the rights of affected communities.

The Ghanaian Mining Sector: A Brief History

The clamor for gold in Ghana (formerly the Gold Coast) is not a novel phenomenon. While West Africa boasts a rich history of artisanal gold mining, European/Western participation is largely responsible for three peaks in Ghana's gold production history. The first and second ended abruptly with the escalation of war in early twentieth-century Europe, with the third and largest boom emerging in the late 1980s and continuing to this day.

In 1957, Ghana's independence ushered in a decidedly socialist political agenda and key sectors of the national economy were reorganized to fall under control of the state. However, the effects of the 1973 OPEC oil crisis threatened to devastate Ghana's national economy and the mining sector was pushed into a state of distress. In 1983, amid an uncertain economic climate, the nation underwent structural adjustment policies (SAP) to secure loans from the World Bank and International Monetary Fund (IMF). The procurement of these loans was conditional on the country's implementation of policies encouraging economic liberalization, as well as a recession of public welfare services (e.g., healthcare, education, sanitation, energy provision, etc.) in favor of privatization. Adjustment marked the beginning of Ghana's insatiable appetite for foreign direct investment (FDI) that was critical in reshaping its gold mining sector. The subsequent adoption of the Minerals and Mining Law in 1986 (PNDCL 153) loosened environmental and operational regulations in the industry in an effort to entice foreign investors. These changes were largely responsible for the huge amounts of FDI channeled into the mining sector with more than 6.5 billion U.S. dollars pouring into operations and related service companies from 1983 to 1999 alone. In addition, production has increased well more than 600 percent since 1985 and currently more than

237 companies (six of which are the largest multinational mining corporations in the world) are registered as either prospecting or actively mining gold (Agbesinyale, 2003; Akabzaa, 2000).

As the global price of gold skyrockets, multinational mining corporations are enjoying record profits and have become increasingly influential actors in Ghana's economic scene. Civil society groups point to the passing of the Minerals and Mining Act of Ghana in 2006 as a testament to the power of the mining industry. This new law was passed amid controversy, as it further liberalizes the sector and makes it legal for the state to seize land or authorize its occupation for purposes it deems in the public or national interest (e.g., mining). However, many are skeptical of the actual impact of mining on Ghana's national economy. The United Nations Conference on Trade and Development (UNCTAD) reported that in 2003 Ghana retained only 5 percent of its total mineral export value, earning only 46.7 million U.S. dollars out of a staggering 893.6 million (UNCTAD, 2005). These statistics are similar to a World Bank report published in 2003, which calculated the industry's contribution to total government tax revenues in 2001 to be only 4 percent or approximately 31 million U.S. dollars, despite their combined turnover being in excess of 600 million U.S. dollars. The World Bank's report blamed generous corporate tax waivers (an incentive to encourage multinational investment) for this disparity and found that local communities affected by large-scale mining have seen little benefit to date in the form of improved infrastructure or service provision. The report recommended that broader cost-benefit analyses of large-scale mining (including the social and environmental costs) and consultations with the affected communities need to be undertaken before granting future production licenses (World Bank, 2003).

The human cost of such intense mining activity has been enormous. It has been estimated that more than thirty thousand people from fourteen villages in the Wassa West District (Western Region of Ghana) were displaced to accommodate large-scale mining projects between 1990 and 1998 (Hilson & Nyame, 2006). In January 2006, more than 9,300 landowners in the Brong-Ahafo Region were displaced by Newmont Ghana Gold Limited (NGGL) in the first phase of their Ahafo Mine project, and it was anticipated that more than 20,000 would be evicted by the end of the second phase (Wassa Association of Communities Affected by Mining [WACAM], 2007). Unfortunately, many of those displaced have not been paid adequate compensation for losing their farms, forests, and means of livelihood due to the tremendous ecological damage and negative impacts caused by nearby mining activity.

Impact of Mining on Surrounding Communities

It is estimated that approximately 30 percent of the land surface of Ghana has been committed to surface mining concession (WACAM, 2007) and while mining projects are scattered throughout the country, the Western Region is widely recognized as the most heavily mineralized and mined area in the country and perhaps in all of Africa. Surface mining technology is used by all multinational gold companies in Ghana. It excavates huge tracts of soil and uses cyanide to extract deposits of gold ore. This practice is a constant source of concern among community members, as more than four major cyanide spills having occurred in the last decade. Though celebrated for its cost-effectiveness, this technique is also extremely invasive and has many implications on the surrounding environment. Current surface mining operations pose a particular threat to nearby farm and forest-based populations, as the vast majority of rural dwellers in Ghana derive their food and income directly from their lush natural surroundings. Ironically, farming is the cornerstone of the Ghanaian economy with agriculture accounting for approximately 36 percent of Ghana's gross domestic product (GDP), in contrast to large-scale mining that is said to only contribute approximately 2 percent (Agbesinyale, 2003; Akabzaa, 2000).

Mining activity has also had a significant impact on local access to potable water, as the vast majority of the rural population depends solely on local rivers and streams as a water source. A study conducted by the Environmental Chemistry Division of Water Research Institute of the Council for Scientific and Industrial Research (CSIR) sampled sources of drinking water in the Wassa West District and discovered dangerously elevated pollutant levels, such as fecal coliform, suspended solids, manganese, iron, chromium, lead, acids, and mercury (1998, as cited in Akabzaa, 2000). Recently, the surrounding communities accused NGGL of intentionally disposing fecal waste from their sewage treatment area by hidden pipe into a major tributary near their Ahafo mine in Kenyase. NGGL denied that the contamination was intentional and claimed that the event was caused by an overflow caused by rain; however, many remain unconvinced (Ghana News Agency, 2005). Water pollution is also caused by the illegal and unregulated activity of small-scale miners (*galamsey*). Their operations have contributed significantly to land degradation, water contamination, diversions, siltation, erosion, and the degradation of streams through bed excavations, panning and soil washing, as well as elevated levels of mercury that is commonly used to process gold ore (Agbesinyale, 2003; Hilson, 2002).

In addition, communities have endured innumerable social impacts due to accelerated mining practices, which Agbesinyale (2003) documents in his detailed case study of mining in the Wassa West District. These include increased rates of child labor, school-dropout, cost of living, general poverty, unemployment, health-related issues (e.g., respiratory, skin, gastrointestinal, and eye/ear disorders, malaria, sexually transmitted diseases, etc.), as well as air, water, and noise pollution. Increased land alienation due to surface mining and the exclusion of youth from compensation and relocation packages are thought to be responsible for overcrowding, loss of housing, increased drug use, and increased incidence of crime in the swelling urban centers. Youth are also flocking to work in nearby *galamsey* camps,[2] as farming and traditional means of employment become less and less sustainable. This decreased vocational interest in agricultural production has helped to foster a growing estrangement from local food production in mining communities, which is paradoxical, as the nation's most fertile area now finds itself heavily dependent on expensive imported foodstuffs. In fact, food security was identified as a primary issue of concern among women in particular. However, their lack of land and resource control, as well as their exclusion from major decision-making processes was identified as the cause for these concerns being largely overlooked, as men were reportedly more concerned with issues of compensation and royalty payments.

The considerable social and environmental impact of surface mining has caused widespread exasperation and infuriation in affected communities. Frustrations mount as corporate promises of prosperity, compensation, jobs, and a better quality of life fail to come to fruition. Disappointed and exploited, communities are forced to witness the deterioration of their surrounding forests, lands, and water ways, as well as their own concomitant dislocation from the earth that has sustained them throughout generations. As a result, popular resistance to large-scale mining in Ghana is gaining momentum and communities have partnered with organizations in a PAR project that builds solidarity and helps them advocate more effectively for their rights. Such responses are consistent with a wider phenomenon of resistance from the grassroots seen around the world, which is indicative of the collision between the place-based interests of these communities and the expansive, destructive trajectory of mining and similar megadevelopment projects.

WACAM and PAR: Research, Organization, Community Activism, and Learning

Community frustration around issues of resource control, the unfair distribution of mining revenue, land alienation, socioecological dislocation, compensation, and resettlement catalyzed the creation of a community-based organization called the Wassa Association of Communities Affected by Mining (WACAM). Established in October 1998, this organization has grown to include approximately sixty communities and tens of thousands of members. It is managed by a nine-member executive council, seven of whom are from affected communities. WACAM is primarily concerned with the malpractice of foreign multinational mining companies and seeks to encourage and enrich emerging activism in affected communities by equipping communities with the skills and knowledge needed to "network for the protection of the environment, natural resources and the rights of marginalized mining communities through advocacy, campaign and representation within a legal framework that is sensitive to the concerns of mining communities" (WACAM, 2007, p. 5).

Despite their initial focus in the Wassa West District, WACAM has grown to include mining communities from across the country and has evolved into an emblem of advocacy for affected communities. WACAM now operates in four general areas (Tarkwa, Obuasi, Kenyase, and Akyem) within four regions of Ghana (Western, Ashanti, Brong Ahafo, and Eastern Regions). Yet despite their expanding scope and increased exposure, they refuse to collect any membership fees from the communities and scrutinize funding sources carefully, so as not to compromise the politics or direction of their projects. This limits their operations considerably and many initiatives are unfortunately delayed or just not possible due to lack of funding. They remunerate only four employees, several of whom also maintain full-time employment elsewhere to make ends meet. As a result, WACAM relies heavily on volunteerism from all members, whether urban-based intelligentsia or village farmers.

Workshops, training programs, and community sensitization projects are the most visible initiatives undertaken by the organization. Community representatives are selected from mining communities to participate in themed workshops to gain a better understanding around a variety of topics, which deepen their ecological, legal, and technical expertise. It is anticipated that these representatives will return to their communities with an increased understanding around certain key issues and disseminate that knowledge that will gradually strengthen communities' ability to negotiate and make demands on corporations and government. Community rights education is central to these projects and the organization employs a rights-based approach as an advocacy strategy given the growing significance of rights-based activism in international/national locations.

However, workshops also serve as a forum for a type of learning described in Foley (1999) that is deeply embedded within participation in struggle. This type of learning builds on Freire's (1970) conceptualization of conscientization, or a process of learning that "challenges and significantly alters a participant's understanding of the world" (p. 39). It is this kind of learning that helps participants understand their engagement in struggle as a transformative process toward addressing their multiple marginalizations. Creative group exercises are used in workshops to reflect and discuss the meaning of concepts, such as development, progress, human rights, and democracy. In fact, transcripts from these sessions show the beginnings of a collective unravelling of commonly held assumptions concerning these issues and the unmasking of contradictions in local perceptions of reality. For example, in May 2003, WACAM began a series of workshops targeting Assemblywomen and community leaders from mining communities in the Atuabu, Abekuase/Samahu, Damang, Bogoso, Nkwantakrom, and Ayensukrom zones. Although these sessions were intended as a leadership training exercise, reports describe women engaging in extensive reflection-based activities around their role in society and in the local economy. Through this reflexive engagement, the women came to recognize their contributions as providers of unwaged labor and the importance of this work in maintaining existing social structures (Owusu-Koranteng, 2003).

WACAM workshops are also profoundly valuable in that they allow communities from around the country to meet and discuss their shared experiences with mining—an opportunity they probably would not have otherwise. For example, a project conducted by WACAM in February 2003 gathered participants from different communities surrounding the Bogoso Goldfields Limited mining operations. They were asked to engage in a group exercise where they compared their quality of life before and after mining projects began in the area. Despite their acknowledgment of the company's acts of "good-will" (i.e., donation of foodstuffs and drilling of several boreholes), there was an overwhelming agreement by all in attendance that their collective quality of life had deteriorated significantly due to mining activity. The following excerpt from a workshop transcript (WACAM, 2003) is testament to this.

> Our community was lively; people had money to care for every member of the family. Our children were in school and men were working... Our community had nutritious foods and we were not getting sick often. We had money and the good things that nature could provide... Our forest had Mahogany, and herbs. We were not getting sick frequently because we were having fish and meat for our meals from the forest... Before surface mining, we had potable water from rivers Twigyaa, Monkoro, and Boadie... Our whole lives and families are now disintegrated. We have no homes and live in great hardship. We were forced to abandon our farms and properties to relocate to other communities. We have become poor out of this. (pp. 3–9)

WACAM also invites communities who have endured the effects of mining activity to meet with communities from prospective mining areas. In June 2005, communities under exploration by Randgold Corporation in the Adansi West and Asutifi North were invited to attend a WACAM workshop for communities in the Wassa West District, where heavy surface mining activity has been ongoing for decades (Owusu-Koranteng, 2005). This allowed prospective mining communities to hear firsthand accounts of the impact of mining from town and villages that had been living with it for decades.

Outside of workshops, WACAM-community participation in the organization and execution of protests, petitions, rallies, community meetings, and demonstrations deepens related organizational and administrative skills of activists and helps them understand that direct action is effective, flexible, and creative. These events also act as platforms for the exchange of speeches, slogans, and symbols that are inherently educational. They embody the spirit of the movement and help propagandize the goals and demands of the community toward the formation of a resistance identity (Routledge, 2003). In fact, WACAM's work over the past decade has contributed to an emerging culture of resistance in Ghana. More groups are beginning to organize and mobilize in the face of destructive development initiatives (i.e., mining, dams, forestry, etc.) using a nonviolent approach that is focused on research and community education. For example, a relatively new organization called the Concerned Farmers Association of Teberebie, whose members and leaders have worked closely with WACAM in the past, have also become vocal advocates for the rights of mining communities. A newspaper article describes the organization's reaction to cyanide contamination into two local streams believed to have been caused by nearby mining activity:

> Cyanide spillage has been found in two streams that serve as sources of water for peasant farmers of Teberebie in the Wassa West district. The two streams that have been reportedly polluted by the poisonous chemical are located at the south gate of Goldfields Ghana Ltd. and behind the waste pile of AngloGold Ashanti Iduaprim Mines. This has made it difficult to determine the exact source of the pollutant...Emelia Amoateng, leader of the Concerned Farmers of Teberebie, described the harsh conditions under which mining communities, particularly those within the Teberebie community were living. "All our water bodies: our heritage have been destroyed by these companies, and they have refused to provide us an alternative; why?" She asked. (*The Heritage,* 2007)

Although WACAM often enters into a community partnership as activist educators, their role quickly evolves to one that is more supportive in capacity. In fact, WACAM clearly announces their "vision to be ultimately transformed into a social movement well structured with resources and capacity to influence policies in favour of the marginalised people, especially those living in mining communities" (WACAM, 2008, p. 7). The organization then becomes more of a facilitator and provides technical, logistical, and moral support, encouraging local activists to emerge within a community and take charge in directing the course of struggle and action.

There are innumerable outcomes stemming from these various WACAM-community PAR projects. Generally, community members have become more attuned to underlying corporate agendas and have become more careful and confident in their negotiations with companies. WACAM-community research has also revealed the importance of using media outlets strategically, and it emphasizes the efficacy of nonviolent initiatives in attracting public attention and sympathy. This PAR project has also had an important role in the cultivation of local activist leaders.

In fact, many of the most vocal community-based advocates from various communities around the nation have roots in WACAM-sponsored workshops or projects. Yet, perhaps most importantly, these PAR projects have helped community activists better understand the connection between resistance, action, and social change. As they develop this sense of agency, mobilize, and are able to assert themselves toward small yet significant achievements, their faith in a process of struggle as a means to affect change is consistently renewed. Although these realizations and victories may seem isolated and infinitesimal, they accumulate as small steps toward tangible and significant change from a community's point of view. For example, after decades of struggle around the globe, it was announced in April 2007 that 92 percent of Newmont Gold Corporation shareholders voted to require the company to address community opposition to its operations around the globe, as investors expressed their concern over patterns of resistance to projects in Indonesia, Ghana, Peru, and the United States (Graman, 2007).

However, the very nature of these WACAM-community PAR projects insists on constant reflection and begs activists to scrutinize the intentions and consequences of their own practices. Aware of their perceived position of heightened expertise and authority by communities, activists need to consider the sources of knowledge informing the struggle, as well as recognize the vulnerability of these partnerships to intrastruggle dynamics of power and their potential in creating new relationships of marginalization. As a struggle unfolds and activists become increasingly influential in illuminating possible directions for struggle, hierarchies are poised to develop and silently stratify a community. Is it possible that some community members are hesitant to support the avenues of resistance suggested by the increasingly influential organization-community coalition? Or, are these concerns marginalized in favor of pursuits that coalesce with the direction endorsed by a more dominant activist group? For example, present activism in mining communities tends to focus on more reactionary forms of advocacy (e.g., compensation, resettlement, etc.) and while mining communities undoubtedly deserve these reparations, there is glaring lack of energy being invested into initiatives demanding greater community inclusion at a decision-making level (e.g., resource control, ownership, granting of licenses, etc.). Is this emphasis on compensation truly representative of the entire community? An authentic PAR project should be aware of an activist's invariable influence on a community's understanding of struggle, as well as the implications of these engagements on the direction of a movement, whose knowledge and/or ways of understanding are legitimized or in turn silenced through these partnered learning engagements. Activists also need to recognize their unique position in opening up possibilities for new paths in struggle, ones which are decidedly political, historical, and collective. Recognizing the centrality of learning in shaping and directing struggle, more specifically in their transformative potential of learning as conscientization, reveals the opportunity for a process of repoliticization of previously silent/silenced communities. These community activists thus play an important role in the encouragement of local translations of ideological suspicions of development paradigms into directed action and in ensuring that a more authentic PAR project is realized.

Concluding Reflections

The globalization of industrial development, in the form of large-scale multinational-led projects, continues to meet resistance by those who depend on the sanctity of their surrounding environments to survive. These struggles to reclaim development occur in the face of overarching, interrelated, structural barriers, which are characteristic of subaltern realities in the South and tilt the terrain of struggle to the disadvantage of these marginalized groups. Communities are already consumed by daily struggles with poverty-based issues, such as, illiteracy, the exaggerated subordination of women, child labor, unemployment, threatened food security, and water scarcity. These hardships are then exacerbated by a community's dislocation from surrounding ecosystems and the social impacts of destructive megadevelopment projects, making participation in resistance movements that much more difficult. In addition, these struggles are situated within a globalized capitalist economy that prioritizes profit over community concerns around issues of health, education, and ecological sanctity. SAP and other neoliberal policies directly affect life in communities by forcing the endorsement of legislation conducive to increased FDI and a strong multinational presence. This effectively subordinates community demands for measures to temper the socioecological impacts of megadevelopment, which threaten to narrow corporate profit margins and investor confidence in projects deemed to be in the interest of national development.

Despite political freedom from colonial rule, subaltern (peasant, tribal, indigenous, rural women, and farmers) groups continue to endure new forms of marginalization within their own sovereign societies, as they are forced to conform to the interests of national development through rapid industrialization and economic growth (Kapoor, 2008; Kwai Pun, 2008). In other words, the dislocation and displacement of subalterns demonstrates the exclusion of subaltern groups from the national identity and conscience, as development is often conducted at their expense. Parajuli (1998) describes this as an ethnicized project, explaining that "ecological ethnicities have to bear a disproportionately large share of the burden of displacement and other negative consequences of development programmes" (p. 188). Popular disenchantment with incumbent development practices has encouraged collective resistance by subalterns in the South, as illustrated in this case study of WACAM-community PAR projects struggling with large-scale mining. However, the rejection of the wider development project does not necessarily imply a desire to turn back time or return to nature, nor does it mean that these groups are not interested in improving their quality of life. Rather, these groups seek to evolve understandings of development toward a practice that privileges local grassroots initiatives over grandiose global action, promoting simpler and less materially intensive ways of living (Peet & Hartwick, 1999).

Community-based PAR is thus invaluable in that it is a profoundly educative process that draws on community research/knowledge generation aimed at addressing the structures of domination felt by those being victimized by such destructive practices. It supports and encourages communities to convene, discuss, mobilize, study, observe, analyze, raise awareness, realize, petition, protest, organize, and plan. In fact, it is through these often incidental and informal educational engagements that community activists and communities move to a position where they are better able to politicize the process of development and situate themselves and their struggle historically and within the constraints and contradictions of today's globalized capitalist economy.

Notes

1. PE understands teaching/learning as an inherently political process, one where neither the teacher nor the learner is neutral. It seeks to address issues of oppression and struggle, recognizes the relationship between knowledge and power, and realizes the role of adult education/learning in confronting these oppressive relationships that typify subaltern struggles in the South (Kapoor, 2008). PE challenges an authoritarian model of education (banking education) and seeks to transform teaching/learning into a directed, but democratic, participatory, dialogical, problem-posing process, which is based on people's experiences, understanding, and knowledge, in an effort to transform the existent social scaffolds of power through the persistent interrogation of the status quo and hegemonic assumptions that shape a person's perception of reality.
2. *Galamsey* camps are notoriously dangerous. These rudimentary operations are responsible for dozens of deaths each year and are prone to collapse. In fact, operators generally work without the most basic of safety measures (e.g., helmets, gloves, boots, etc.).

References

Agbesinyale, P. (2003). *Ghana's gold rush and regional development: The case of the Wassa West District.* Dortmand, Germany: University of Dortmand.

Akabzaa, T. (2000). *Boom and dislocation: The environmental and social impacts of mining in the Wassa West District of Ghana.* Accra, Ghana: The Third World Network—Africa.

Fals-Borda, O. (1988). *Knowledge and people's power.* New Delhi: Indian Social Institute.

Foley, G. (1999). *Learning in social action: A contribution to understanding informal education.* New York: Zed.

Freire, P. (1970). *Pedagogy of the oppressed.* New York: Continuum.

Ghana News Agency. (2005, December 13). *WACAM condemns disposal faecal matter into Asuopre stream.* Retrieved from http://www.ghanaweb.com/GhanaHomePage/NewsArchive/artikel.php?ID=96004

Graman, K. (2007, April 25). Newmont mining shareholders order environmental reform. *The Spokesman-Review.* Retrieved from http://www.spokesmanreview.com/local/story.asp?ID=186358&page=1

Hilson, G. (2002). The environmental impact of small-scale gold mining in Ghana: Identifying problems and possible solutions. *Geographical Journal, 168*(1), 57–72.

Hilson, G., & Nyame, F. (2006). Gold mining in Ghana's forest reserves: A report on the current debate. *Area, 38*(2), 175–185.

Kapoor, D. (2007). Subaltern social movement learning and the decolonization of space in India. *International Education, 37*(1), 10–41.

———. (2008). Globalization, dispossession and subaltern social movement (SSM) learning in the South. In A. Abdi, & D. Kapoor (Eds.), *Global perspectives on adult education* (pp. 100–132). New York: Palgrave Macmillan.

Kwai Pun, V. (2008). *Mining displacement and learning in struggle in Ghana*. Unpublished Master's thesis, McGill University, Montreal, Canada.

———. (2008). Popular education and organized response to gold mining in Ghana. In A. Abdi, & D. Kapoor (Eds.), *Global perspectives on adult education* (pp. 253–275). New York: Palgrave Macmillan.

Owusu-Koranteng, H. (2003, May). *Women leadership workshop for women in mining communities—Module One*. Tarkwa, Ghana: WACAM.

———. (2005, July). *Report on three-day workshop using the rights based approach in strengthening mining community capacity*. Tarkwa, Ghana: WACAM.

Parajuli, P. (1998). Beyond capitalized nature: Ecological ethnicity as an arena of conflict in the regime of globalization. *Ecumene, 5*(2), 186–217.

Peet, R., & Hartwick, E. (1999). *Theories of development*. New York: Guilford.

Rahnema, M. (1992). Participation. In W. Sachs (Ed.), *The development dictionary: A guide to knowledge as power* (pp. 116–131). London: Zed.

Routledge, P. (2003). Voices of the dammed: Discursive resistance amidst erasure in the Narmada Valley, India. *Political Geography, 22*(3), 243–270.

Smith, L. T. (1999). *Decolonizing methodologies: Research and indigenous peoples*. London: Zed.

The Heritage. (2007, September 19). *Mining companies pollute streams in Teberebie*. Retrieved from http://news.myjoyonline.com/health/200709/8772.asp

The World Bank. (2003, July 1). *Ghana - Mining sector rehabilitation project and the mining sector development and environment project*. Retrieved from http://www-wds.worldbank.org/external/default/main?pagePK=64193027&piPK=64187937&theSitePK=523679&menuPK=64187510&searchMenuPK=64187283&theSitePK=523679&entityID=000094946_03081404004344&searchMenuPK=64187283&theSitePK=523679

United Nations Conference on Trade and Development [UNCTAD]. (2005, September 13). *Economic development in Africa: Rethinking the role of foreign direct investment*. Retrieved from http://www.unctad.org/en/docs/gdsafrica20051_en.pdf

Wassa Association of Communities Affected by Mining [WACAM]. (2003, February). *Baseline studies within communities affected by Bogoso Goldfields Limited*. Tarkwa, Ghana: WACAM.

———. (2007, January). *Annual Report 2006*. Tarkwa, Ghana: WACAM.

———. (2008, March). *Annual Report 2007*. Tarkwa, Ghana: WACAM.

CHAPTER THIRTEEN

Ethnography-in-Motion: Neoliberalism and the Shack Dwellers Movement in South Africa

Shannon Walsh

This chapter unravels the role of ethnography as an aspect of political practice in the context of shack dweller struggles in postapartheid South Africa. Using both macro and micro optics, it is possible to examine globalization under late capitalism and its specific conditions and variations manifest in a South African shack settlement in Crossmoor, Chatsworth, with a discussion of how an ethnography-in-motion has been used as a pedagogical and political methodology. I follow Gillian Hart (2002) in hoping to "clarify the slippages, openings, and possibilities for emancipatory social change in this era of neo-liberal capitalisms, as well as the limits and constraints operating at different levels" (p. 45).

In response to the widening gap between the rich and poor in postapartheid South Africa, there has been a blossoming of new social movements. In 2006 alone there were almost eleven thousand protests recorded throughout the country, a majority of which were characterized as service-delivery protests with demands around basic services such as water, electricity, sanitation, land, and houses. At the core of these new social movements are the more than 3 million people who live in shacks (Statistics South Africa, 2006), and millions more who live in inadequate, overcrowded flats and crumbling township houses.

While shack dwellers have been radically impacted at the level of housing, water, and sanitation, it is clear that neoliberal economic policies are also directly linked to the rise of AIDS in shack settlements that tower to nearly 70 percent in the communities I discuss here. In response to these conditions, practices, pedagogies, and resistances are emerging from those on the front lines.

Given this context, this chapter explores how an ethnography-in-motion has been used within a Durban shack settlement as a practice for militant research. Ethnography-in-motion is grounded in knowledge arising from social movements and from the everyday practices of people living in, through, and against neoliberalism. It is also a practice that includes engaged pedagogy, visual methods, and other creative means of organizing and working collaboratively to create concrete engagements in social struggles.

Rather than a prescription to be followed, ethnography-in-motion is a proposal with movement and fluidity, an attempt to situate the doing of ethnography within a political practice. Rather than trying to inscribe a new piece of jargon into an already cluttered intellectual lexicon, I hope that by thinking through ethnography creatively, the doing of ethnography can be situated within a political practice. It is an attempt to elaborate in a militant research space what the Zapatistas mean when they say, "Walking, we ask questions."

Notes from the Margins

Neoliberalism in Postapartheid South Africa

Neoliberalism has come about in very particular ways in South Africa that macro theories can often oversimplify. Yet forming over the ashes and foundations of the apartheid system, the rush toward neoliberal economics has taken a brutal form, what Patrick Bond calls "the world's most extreme site of uneven capitalist development" (Bond, 2001, p. 31). Many see the adoption of the Growth, Employment, and Redistribution (GEAR) policy framework in 1996 as part of continuing class division and capitalist relations long existing in the country (Bond, 2001, 2004; Hart, 2002; Marais, 2001; Wolpe, 1972),[1] solidifying a neoliberal program for postapartheid South Africa. Even after more than a decade of betrayals, many people, especially the poor black majority, are still coming to terms with the fact that the promise of socialism—the rather abstract Freedom for All—that was the hallmark of the African National Congress (ANC) platform during apartheid, looks more like a fantasy than ever.

Bond and Desai (2006) describe what has emerged in South Africa as uneven and combined development, and follow David Harvey (2006) in analyzing what they see as accumulation by dispossession, or as Bond calls it looting Africa. By the early 2000s, "with one of the world's most entrenched systems of urban inequality, South Africa fell from 86th to 120th place out of 177 countries on the Human Development Index" (Bond, 2006, p. 19). Given South Africa's relatively high Gross Domestic Product (GDP), its low rating on the Human Development Index (HDI) (a scale based on literacy, education, life expectancy, and standard of living) is particularly acute. The growing gap between the rich and poor is a politically sensitive issue since it has become wider since the days of apartheid and "[b]etween 1975 and 1995, HDI improved fairly steadily as life expectancy and educational enrolments improved. Since democracy arrived, HDI has fallen, hitting a new low last year" (Butler, 2007). While the HDI^2 has flaws as an accurate marker for overall well-being, it is nonetheless an indicator of growing disparities in an otherwise wealthy country. Recent Gini coefficient data, which takes into account social and economic disparities, has also shown that unequal wealth distribution in South Africa has grown considerably from 1996 to 2005 (Ntshalintshali, 2006).

Clearly, AIDS has had a significant impact on the HDI and the Gini coefficient. The pandemic has brought life expectancy in South Africa down from sixty-two years in 1990 to forty-seven years in 2005 (United Nations Population Fund [UNFPA], 2005). AIDS-related deaths in South Africa had taken more than 2 million people by the end of 2005. Between 1997 and 2005, the annual number of registered deaths for people between twenty-four and forty-nine years old rose by a staggering 169 percent (Medical Research Council [MRC], 2006). There are more than 5.5 million South Africans living with HIV, including 240,000 children less than fifteen years old, and more than a million children orphaned (United Nations Programme on HIV/AIDS [UNAIDS], 2007). The province of KwaZulu-Natal (KZN) has one of the highest rates of infection in the country at 39 percent in 2006 (South African Department of Health, 2006). Yet by 2007, only 300,000 people had accessed antiretroviral (ARV) medication from the government, though there are plans to reach one million people by 2010 (South African National AIDS Council [SANAC], 2006; UNAIDS, 2007). Colleen O'Manique (2004) argues that neoliberalism's entrance into the global sphere is at the heart of the dilemma posed by AIDS. She insists,

> the problem is not simply that clinical medicine fails to pay sufficient attention, but that it is difficult for our highly individualistic societies to see the link between the broader organization of societies, the distribution of wealth and power, and human health. Unfettered growth and trade, understood as "freedom", overrides other values. (p. 19)

As the gap between the rich and the poor widens, the impacts of AIDS worsen. Mark Hunter (2005) has highlighted clear links between economic policy and AIDS in South Africa, specifically as it manifests in informal settlements "where HIV rates are reported to be almost twice as high as they are in rural and urban areas" (p. 145). The higher infection rates in shack settlements are "not only due to sexual transmission, but is in part a consequence of inadequate water, nutrition, and sanitation and the general poor state of health in the former" (p. 145). AIDS is rooted to socioeconomic forces in far more complex ways than is often assumed (Hunter, 2005).

While neoliberalism "has guided the globalization of economic activity and become the conventional wisdom in international agencies and institutions (such as the IMF, World Bank, World Trade Organization, and the technical agencies of the United Nations, including the WHO)" (Navarro, 2006), it is increasingly being challenged as adverse effects on people and societies pile up. While the specific adaptations of neoliberalism in postapartheid South Africa is not the sole cause of the disparities we are witnessing, an examination of these policies does "capture the unwillingness of the state to intervene more directly to redistribute wealth, create employment, and provide basic services" (Hunter, 2005, p. 151).

In turn, the adoption of neoliberal policies in so many areas that directly impact people's ability to sustain life itself, combined with the increasing AIDS crisis, can arguably be directly linked to the rise of new social movements in postapartheid South Africa. For this reason, it is crucial to look to the issues as identified and articulated by social movements in response to the intrusion of privatization and commodification into their lives. I follow Stevphen Shukaitis and David Graeber's (2006) insistence that "militant research starts from the understandings, experiences, and relations generated through organizing, as both a method of political action and as a form of knowledge" (p. 12). Particularly, I focus on the shack settlements in the south Durban area of Chatsworth, called Crossmoor, where I have been doing some of my fieldwork.

Crossmoor is comprised of two shack settlements inhabited by approximately four hundred Indian and African families who have been dealing with ongoing evictions, court cases, political promises, and media attention since their first attempts to occupy the land in August 2006. In that time the communities have also successfully mobilized marches, meetings with councilors and other city representatives, started a nonprofit shack dwellers organization, launched campaigns in the form of court cases and letter writing, and managed to formalize many aspects of their community while being consistently under attack.

The research approach I adopted in Crossmoor, what I call ethnography-in-motion, hinges on two interlinking ideas. First, it is a grounding in knowledge arising from social movements and from the everyday practices of people living in, through, and against neoliberalism. Second, it is a practice that includes engaged pedagogy, visual methods, and other creative means of organizing and working collaboratively to create concrete engagements in social struggles. In this way, it brings together a grounding in knowledge produced through direct engagement, and an active, engaged aspect that stimulates further knowledge production and political action. To begin then, I am interested in examining the responses to the impacts of neoliberal economic policy delivered from above from social movements in South Africa, setting into context the situation in Crossmoor as we zoom in more closely.

Social Movement Issues against Privatization

Postapartheid social movements in South Africa have been considerably focused on issues around the privatization of water, electricity, and housing.

In Johannesburg, the Soweto Electricity Crisis Committee (SECC) was formed to respond to the commodification and privatization of electricity, as South Africa began to privatize its formerly state-owned electricity and water utilities. In 1999, South Africa privatized the state-owned electricity provider Eskom, which led to the installation of prepaid electricity meters and massive electricity cutoffs for thousands of people living in poor communities. Instead of bringing efficiency, these new corporate formations focused on full cost recovery, with prices at times increasing by up to 400 percent (Smith, 2002), which had radical impacts on those who could not afford to pay.

At the same time as electricity was being privatized, so was water. The water utility in Johannesburg was bought by Suez, a French consortium, which also quickly led to higher rates, prepaid water meters, and thousands of water cutoffs. Again social movements and spontaneous struggles, mainly focused on prepaid water meters, disconnections, and water cutoffs, erupted across the country. Prominent groups include the Anti-Privatisation Forum (APF), the Coalition against Water Privatisation (CAWP), and the Phiri Concerned Residents Forum (PCRF). Many groups and communities have also been reconnecting water for residents who have been cutoff because of nonpayment. In Phiri (Soweto), the struggle against prepaid water meters has been ongoing since their installation by Johannesburg Water in 2003. In 2008, they will be heard in the high court.

Throughout 2006 and 2007, residents from Crossmoor marched on the city, wrote letters, and told their stories to the media in an attempt to secure water and toilets for their more than three hundred residents who had neither. In Kennedy Road, a shack settlement within the city of Durban that is almost twenty years old and the home to *Abahlali base-Mjondolo* (shack dwellers movement), more than seven thousand people live with only six serviced toilets. Consequences of not having access to water have been deadly for poor communities who often live with up to twenty people in a household, and face regular cutoffs when they are unable to pay. Inadequate amounts of water and proper sanitation are a critical public health issue, deepening the severity and intensity of the AIDS crisis (APF, 2007; Bond, 2006).

Simultaneously, land and housing are major issues in contemporary South Africa. The country continues to be marked by the racist geographic separations used by the apartheid government through legislation, such as The Group Areas Act (1950), which tore apart and ghettoized whole communities based on the color of their skin. This legacy is not easily shorn and is a continual reflection of the legacy of inequality and the unequal distribution of wealth. Housing issues have remained at the center of discussion around how the transition and promised redistribution of wealth can be measured. For the ANC, housing has been a central discursive feature of their political campaigning. Hunter (2005) reports that "[d]uring the democratic transition, the ANC heralded the provision of housing, perhaps more than any other policy, as having the ability to jump start radical economic and social redistribution" (p. 157).

While around a million houses have been built, the reality and approach to providing housing has been far off target in a number of ways. Hunter (2005) writes, "[t]he state took twice as long (ten and not five years) to meet its first target of building one million new houses. Some scholars blame shortcomings in housing policy on the weakness of the chosen mechanism for delivery, namely a market-driven one off capital subsidy system" (p. 157). Postapartheid informal housing planning has been based largely on the Urban Development Framework that took an individualizing and commodifying approach to development, relying primarily on market mechanisms to deal with housing (Huchzermeyer, 2004).

Recently, antipoor legislation of slum clearance, such as KZN province's Elimination and Prevention of Re-emergence of Slums Bill (KZN, 2006), coupled with the increased repression around land and housing in the form of ongoing evictions, shack demolition and forced removals, is making life even more precarious for millions of South Africans. Social movements such as Abahlali baseMjondolo, Western Cape Anti-Eviction Campaign (WCAEC), the Westcliffe Flat Dwellers Association, and others have been tirelessly mobilizing around land and housing issues.

As Surooj, a resident of Crossmoor, explains, "We are going to contest the evictions. What can we do? We are desperate. The conditions are bad; it's not healthy living for us at all." Living in overcrowded, deplorable, violent, and untenable situations coupled with run-down government housing, increasing poverty, and joblessness has created no alternatives for many but to build shacks or face homelessness.

Housing insecurity has a number of impacts on AIDS and HIV and other health issues. It makes it difficult to get access to, and remain on, medications for TB and other AIDS-related illness. Sleeping in the bush, or in overcrowded conditions, greatly impacts health and increases the likelihood of contracting TB. Weakened immune systems are more susceptible to HIV and if HIV is already present in the system, conditions which make it difficult to maintain a healthy diet increase your chance of getting sick and dying. The unfortunate irony is that many of the people who end up in shack settlements are trying to cope with a family member who has HIV or other disabilities, or have come from rural communities to the city in search of treatment and care when they are very ill. The cycle of poor health is then exacerbated by the conditions in the settlements and the close proximities of people living with HIV and TB invariably leads to increases in transmission.

In Crossmoor, there are a disproportionate number of people living with disabilities who claimed to have ended up in a shack in large part because of the economic drain on their household due to illnesses. Likewise, many residents of the Kennedy Road shack settlement were deeply impacted by AIDS and were shuttling family members between rural areas and the city in seek of treatment in care, largely without success. Because of these overlapping and intersecting situations, AIDS becomes ghettoized within shack settlements.

Zooming in

It was so sunny that morning. The kind of sun that streams through the trees and picks up all the little particles of dust on the air. I arrived very early in Crossmoor, the sounds of the hammers from the night before still resounding in the air. People were tired but looking content with their labor. A lot of the single women and their children were still building shacks up the side of the hill. But homes were getting built.

I wanted to find Ramen. His five-year-old daughter, Renee, is magical to me. So full of mystery and wonder about the world. She's lovely; a perfect little round face with huge black eyes. Since I've been coming to the dusty shack settlement in Crossmoor she's been following me everywhere, staring at me yet refusing to talk. Sometimes, the way she stares, it seems like she doesn't really know that she exists in the world, that I can see her. The women tell me that she talks a lot when I'm not around. Sometimes she even stands on the little table in the settlement all by herself and sings Hindi songs and dances. Her silence has become a kind of game between us. She stays steady as a rock, willing me to try to break her. She receives tickles, questions, hugs, entreaties, all with as stoic a face as she can muster, beating me every time. I spin her around and she tries hard to repress her blooming smile, lest I see.

Ramen's shack has become important to me, because Ramen is her father. He builds the roof over Renee's head, not to mention her two younger siblings and her mother. The five of them. And Ramen is such a decent man. A good father. A rare man who stayed with his family even in the midst of such hardships and poverty.

The house he built through the night, his shack, was so lovely. Beautiful little windows that opened, and nice red wallpaper inside he got from the signboard shop where he works part-time. He was exhausted, sprawled out on the little bed with the kids all around him. I hadn't seen him look so happy. His eyes shone through his tiredness. "I've been up all night." He smiles. I smile. There's not much to say. I film the little shack, Renee holding onto the door-frame proudly, staring up at me in her usual, provocative way.

When the police and protection services arrive a few hours later people wearily pick up their hammers and spades and make to the front to the settlement. There was so much energy in the air after that night of labor. The joy of finally having homes. Yet the struggle is not over.

People sang protest songs, *toi-toied,* tried to block the road into the settlement, and sent representatives down to reason with the police. None of it mattered. The sound of the wood supports of shacks being smashed with axes and the shouts of people trying to protect their homes resounded from all sides. The low-level protection services workers looked a little scared. They smashed as fast as they could with their axes and hard hats. It's not unlikely that they live in shacks themselves. People were begging and pleading, trying to protect their homes. It was heartbreaking. Hot angry tears welled in my eyes. Shots rang out. People were yelling, "Are you trying to kill us? Then kill us right here! We will die on this land. We can not live like this any longer."

I joined the voices beseeching to the police to stop their unlawful attack. Coldly they looked away and avoided my questions. "There are children there! What are you doing?" I screamed as an officer aimed his gun toward Ramen's house. "I'm not going to kill anyone, it's only rubber bullets," he shot back at me.

An hour later Ramen's house was in tatters, now only boards lying on the grass. Ramen just kept shaking his head, hung down low, sadly looking at the remains. He was speechless. Renee and her one-year-old sister sat under a little piece of foam, erected as a make-shift shelter to protect them from the burning sun.

Without lifting his head Ramen starts talking to me, "What can I do? There is nothing for me to do. I have to put a roof over my family's head. I will just have to lift the shack up and live again." (*field notes, 2007*)

Crossmoor. In motion. Zooming in closer to see the impacts of neoliberal policies and forced removals gives them a tone that statistics just do not muster. Like institutional ethnography that "seeks to locate the dynamics of a local setting in the complex institutional relations organizing the local dynamics" (Ng, 1984, p. 19, cited in Choudry, 2004, np), ethnography-in-motion vaults between the complex, often multisited, political and economic environments in which the local occurs, while continuing to mobilize within it. The everyday practices of the state, the police, nongovernmental organizations (NGOs), activists, international financial institutions, and others are all part of the production of power and inequalities. As are the practices of the men who come with axes to tear down the shacks.

There are two security guards that keep twenty-four-hour watch over the settlement, lest anyone try their luck at building a new shack on the municipal land. Yet the guards are living close to the people and this continual interaction is having an impact. They don't want to lose their jobs, but they aren't pleased at having to watch people sleep out in the open fields with their children. During the day they sit idly under the shade of a big tree on the upper part of the land. I ask them about their work and they tell me they are security guards contracted by the city to do its dirty work. They don't like it, but not unlike the residents in Crossmoor, they don't have much choice. After domestic work, you are most likely to find a job these days as a security guard. I ask one of the guards where he lives. He looks down into the dust, embarrassed. He doesn't want to tell me. I guess that he may live in a shack himself as so many of the men I've met in the settlements work as contract security guards, though usually guarding the property of the rich. I'm not sure which would be worse. (*field notes, 2007*)

As this vignette relays, it is through everyday practices that actors "(re)create capitalist relations" (Millar, 2007, p. 8) and are not only dominated by them. As Ramen resists, mobilizes, and creates alternatives when the state offers him none, the state's agents in the form of security guards reproduce capitalist relations in maintaining private property rights and simultaneously allow certain practices to emerge and flourish. Drawing on Bourdieu (1990) in her ethnography of the informal economy of dump sites in Brazil, Kathleen Millar (2007) points out that "capitalism, like any other social system, does not exist apart form human action but must be made and re-made daily through the practices of individuals" (p. 9). The security guards' and municipal workers' sympathy for the Crossmoor settlement (coupled with a simultaneous fear of the mass of people in the community turning on them), and the evolving friendships and relations with community leaders opened up a space for a land occupation even under municipal surveillance.

In response to untenable living conditions, residents of the shack settlements in Crossmoor, Cato Crest, and Mayville initiated an umbrella organization called the South African Shack Dwellers Organization (SASDO). SASDO joined the many other social movements that have emerged in the country linking issues around access to adequate housing, water, electricity, and sanitation to the dignity and health of an increasing number of South Africans.

SASDO's Implementation Strategy focuses on supporting "shack-dwellers in crisis which they experience everyday basis" and looking at an "holistic development approach which involves, e.g. Housing Project, Electricity, Water, Sanitation which involves toilets and roads and Political biasness" (SASDO, 2007). Ultimately seeing the links between these issues and the need to mobilize on a broader front they insist that "in South Africa shack-dwellers are the people who are not enjoying the democracy which some of them they fought for. Shack-dwellers in South Africa are the people who are treated different from other citizens, they are treated as a illegal foreigners" (SASDO, 2007).

SASDO articulates not only a desire to fight for the rights of shack dwellers, of which the organization comprised, but also to "deal with councilors behaviour and attitude of removing shack-dwellers in land close to commercial centres allocated solely to developers" (SASDO, 2007). This makes clear the links between capital exploitation in favor of big business over the right to adequate housing and services for the increasing number of people forced into precarious living conditions.

Night after night, police smashing after smashing, the community persisted to build and rebuild shacks, until finally perseverance, sympathetic city workers, and lenient contracted security guards combined to stabilize a community of 280 new shacks on the hills of Crossmoor. This (new) part of the original settlement named itself *Ekupoleni* (the promised land). It had been over a year since the initial land occupation, forced evictions, and court case had taken place. Six months after the shacks remained standing the settlement had a church, a creche, and every bare piece of ground was growing corn, squash, and flowers for sale in the market and for use by the community. They had mobilized continually during this period, marching on councilors and the mayor, speaking to the local police and security guards, and working together with Congressional Budget Offices (CBOs), government departments, and middle-class activists to achieve their goals. SASDO was born in this time period as well. By 2008, the two parts of the community had stand taps, but continued to wait for adequate toilets. A community had been born, and it was nearly self-sufficient.

Yet it is not, of course, a simple or rosy picture. Racism boils very close to the surface as Africans and Indians struggle to live close to each other for the first time. Power struggles within the leadership exist, as do struggles over resources and status that erupt from time to time, especially with the insertion of outsiders such as myself and other independent activists, and those from government, religious groups, media, and NGOs that have material and other resources to offer. The conditions continue to be very hard, food is a problem, water and sanitation is still desperately poor, and vicious rats have infested the settlement, eating people's food and attacking children, because municipal garbage pick-up is so infrequent. Yet Crossmoor is an interesting, lively place, full of promise, agency, laughter, and hope, while continuing to be a site of much hardship.

"Home-Less" in Crossmoor

While a primary aspect of ethnography-in-motion is a grounding in knowledge produced through direct engagement in struggles, the second part includes creative, critical pedagogy as part of reflexive elaborations on organizing. In Crossmoor, this approach created a horizontal and unbounded pedagogy, mixing discussion, research, mobilizing, organizing, and critical reflection.

Over the course of a year in collaboration with the community, I documented and collected video on various aspects of the struggle in Crossmoor, including organizing meetings, various police attempts to smash people's homes, and everyday living and laughing together. A final twenty-five-minute documentary, "Home-less," was then screened within the community and became a center point for sharing knowledge, discussion, reflection, arguments, and troublesome issues from within the residents' own struggles. Racial divisions that emerged throughout the year that divided the settlement were both contested and revealed in the video screening discussion session.

Simply stated, the video acted as a mirror, reflecting back the historical trajectory of this particular struggle—their struggle—in a way that allowed critical engaged reflection and strategizing. The knowledge that had been produced through everyday organizing and fighting with the police (and at times, each other) was reflected back, elaborated, rehashed, and strategized. In one example, months after the first major confrontation with the police, stories began to circulate that the Indians living in the community had not stood at the front lines to confront the city workers who had come with axes to smash the shacks. The video footage, conversely, showed quite the opposite, with Indians and Africans standing together, linking arms and attempting to stop the destruction of their homes. This was an uncomfortable realization in the space of the screening, yet undeniable and earlier racial divisions tangible in the room softened as the community watched their own history play out before their eyes.

Reflecting back this process through an ethnography-in-motion using video documentation captured ways the state reproduces itself through mechanisms of governmentality, yet kept theory grounded in the real constraints and possibilities actually existing in the world. From the housing struggles in South Africa to the anti-IMF summits, there are so often men with sticks, borders to cross, and places we are shut out from. While a shifting, malleable notion of power is hard to reconcile with the violence witnessed by the state and the police apparatus everywhere we manifest alternatives; remembering that those relations are continually creating and recreating themselves allows us to see fissures and cracks in an otherwise solid looking system of constraints.

Conclusion

In the particular context of South Africa, there is ample evidence of a protracted war against the poor fought through police, through the suppression of life saving medications, the swift rise of neoliberalism, and through the symbolic violence enacted by the growing elite against the majority poor (see, Bond, 2000, 2004; Desai, 2002, 2006; Hart, 2002; Marais, 2001; Walsh, 2008). At a local level, the state of exception[3] functions by creating a being-outside, yet belonging on the very bodies of informal workers, shack dwellers, and migrants. The movement inherent in this making and remaking and the interplay between topographies of power (Ferguson, 2006) is the basis for an ethnography-in-motion. As Holloway (2007) writes,

> Our dignities are stones thrown through the glass of capitalist domination. They create holes, but, more than that, they create cracks that run. Movement is essential. Capital is a constant process of filling the cracks, re-absorbing our rebellions, so that our rebellion, to stay alive, must move faster than capital. An autonomous space that does not spread, that does not become a crack, risks being turned into its opposite, an institution. (np)

This is not a recipe for explaining the world, but a method for sharing stories of resistances, constraints, cracks, and exclusions. It is the interplay between the global and local seen together, through the everyday, simultaneous, and continual effecting of one another. Further, this exchange presupposes that we learn and dialogue better in action, pushes us to reflect while doing, to question while walking. This, then, is part of an ethnography-in-motion.

Notes

1. Wolpe describes the apartheid system not only as one of racial exclusion, but also as a system of domination and control intrinsically bent on reproducing and elaborating capitalist relations already present in South Africa.
2. Patrick Bond points out that an accurate picture of the HDI would need to take account of income disparities in South Africa, the low quality and poor location of housing, the high levels of electricity disconnections and prepaid meters, and the uneven application of the Free Basic Electricity policy.
3. As Giorgio Agamben (2005) explores, the state of exception, or iustitium, has been an integral aspect of the juridical system since Roman times, and the voluntary creation of a permanent state of emergency has become the dominant paradigm for "contemporary nation states, including so-called democratic ones" (p. 2).

References

Agamben, G. (2005). *The state of exception*. Chicago: University of Chicago Press.
Anti-Privatisation Forum [APF]. (2007). *Comrade Nkosingiphile Mvalo Mhlope—Community activist in Alexandra assasinated on Saturday night by unidentified assailants*. Press Release, September 4, 2007. Alexandra Vukuzenzele Crisis Committee.
Bond, P. (2000). *Elite transition: From apartheid to neoliberalism in South Africa*. London: Pluto Press.
———. (2001). *Against global apartheid: South Africa meets the World Bank, IMF and International Finance*. Landsdowne: University of Cape Town Press.
———. (2004). *Talk left, walk right: South Africa's frustrated global reforms*. Scottsville, SA: University of KwaZulu-Natal Press.
———. (2006). *Looting Africa: The economics of exploitation*. Pietermaritzburg: University of KwaZulu-Natal Press.
Bond, P., & Desai, A. (2006). Explaining uneven and combined development in South Africa. In B. Dunn, & H. Radice (Eds.), *Permanent revolution: Results and prospects 100 years on* (pp. 230–244). London: Pluto Press.
Bourdieu, P. (1990). *The logic of practice*. Cambridge: Polity Press.
Butler, A. (2007, November 12). Staring at South Africa's sad human face. *Business Day*. Choudry, A. (2004). Institutional ethnography, political activist ethnography and "anti-globalization" movements. Unpublished paper.
Desai, A. (2002). *We are the poors: Community Struggles in Post-Apartheid South Africa*. New York: Monthly Review Press.
———. (2006). Vans, autos, kombis and the drivers of social movements. Paper presented at the Harold Wolpe Memorial lecture, Centre for Civil Society, International Convention Centre, July 28, 2006.
Ferguson, J. (2006). *Global shadows: Africa in the neoliberal world order*. Durham, NC: Duke University Press.
Hart, G. (2002). *Disabling globalization: Places of power in post-apartheid South Africa*. Pietermaritzburg, SA: University of Natal Press.
Harvey, D. (2006). *Spaces of global capitalism: Towards a theory of uneven geographical development*. London & New York: Verso.
Holloway, J. (2007). *What is revolution? A million bee stings, a million dignities*. Retrieved from http://www.openspaceforum.net/twiki/tiki-read_article.php?articleId=503

Huchzermeyer, M. (2004). *Unlawful occupation: Informal settlements and urban policy in South Africa and Brazil.* Trenton: Africa World Press.

Hunter, M. (2005). Informal settlements as spaces of health inequality: The changing economic and spatial roots of the AIDS pandemic, from apartheid to neoliberalism. *Centre for Civil Society Research Report,* 44, 143–166. Durban, SA: Centre for Civil Society.

KwaZulu-Natal Department of Housing [KZN]. (2006). KwaZulu-Natal elimination and prevention of re-emergence of slums bill.

Marais, H. (2001). *South Africa: Limits to change: The political economy of the transition.* Cape Town: University of Cape Town Press.

Medical Research Council. (2006). *MRC Research Report 2006.* Tyberg, SA: MRC.

Millar, K. (2007). *The Informal economy: Condition and critique of advanced capitalism.* Paper presented at Centre for Civil Society seminar, Durban, South Africa, August 2007.

Navarro, V. (2006). Worldwide class struggle: Neoliberalism as a class practice. *Monthly Review,* 58(4). Retrieved from http://www.monthlyreview.org/0906navarro.php

Ntshalintshali, B. (2006, November 15). Slaves to poverty. *Sowetan.*

O'Manique, C. (2004). *Neoliberalism and AIDS crisis in sub-Saharan Africa: Globalization's pandemic.* New York: Palgrave MacMillan.

Smith, C. (2002, August 30). Guerrilla technicians challenge the privatization of South Africa's public resources. *These Times: Independent News and Views.* Retrieved from http://www.inthesetimes.com/issue/26/22/feature3.shtml

Statistics South Africa. (2006). *Mortality and causes of death in South Africa, 2005.* Pretoria: Republic of South Africa.

South African Department of Health. (2006). *The South African department of health study, 2006.* Pretoria: Republic of South Africa.

South African National AIDS Council [SANAC]. (2006). *National strategic plan 2007–2011.* Pretoria: Republic of South Africa.

South African Shack Dwellers Organization [SASDO]. (2007). *SASDO constitution.* Durban, South Africa.

Shukaitis, S., & Graeber, D. (Eds.). (2006). *Constituent imagination: Militant investigations/collective theorizing.* Oakland, CA: AK Press.

United Nations Population Fund [UNFPA]. (2005). UNFPA in South Africa. Retrieved from http://www.unfpa.org/hiv/gyp/profiles/southafrica.htm

United Nations Programme on HIV/AIDS [UNAIDS]. (2007). *Country situation analysis: South Africa.* New York: UNAIDS.

Walsh, S. (2008). Uncomfortable collaborations: Contesting constructions of the poor in South Africa. *Review of African Political Economy,* 35(2), 255–279.

Wolpe, H. (1972). Capitalism and cheap labour power in South Africa: From segregation to apartheid. *Economy and Society,* 1, 425–456.

CHAPTER FOURTEEN

Kabyle Community Participatory Action Research (CPAR) in Algeria: Reflections on Research, Amazigh *Identity, and Schooling*

TAIEB BELKACEM

As Charles Péguy, a French socialist writer and Catholic who died on the battle field during the First World War, would say, "every mystic starts in and finishes in politics..." (Péguy, 1910/1993, pp. 102–103). I met the Innu nation in Canada in 1999 and started my healing journey within the Midiwin society of North America. At first I felt unbalanced by the strong quest for identity within the Innu nation but then I started coming back to my roots as a Berber and began to stand on my own two feet again. I was much more comfortable being in charge of my own destiny and felt more empowered by the idea of freedom that is intrinsic to my *Amazigh* (free man) identity.

What I found interesting while living with my Innu family was the organization of the community around a system of beliefs different from the Western system I knew before and that I considered until then as the only one. This ethnocentrism became a shame for me as I realized that my vision was blurred and the cultural empathy that I claimed to have was also in some way portraying my ignorance. I realized I was simply the product of a cultural demagogy and ethnocentric school system in France, which prevented me from being aware of myself beyond the self. It is this thought in part, that brings me to this project of community participatory action research (CPAR) in my village.

As I have been taught in the Midiwin society, to heal people you first need to heal yourself. In the healing process education plays a big part; the acquisition and protection of my indigeneity and our knowledge system is another significant element. This knowledge that I am attempting to bring back to the fore has been hidden for many years since it has been treated by others as useless, retarded, shameful, and outside the realm of the hegemonic scientific field of knowing. Getting involved in my village life and participating in educating myself as well as the people of my village has brought on the possibility of social change and participatory work for the protection of *Amazigh* indigenous knowledge.

Talking about CPAR and Kabyle Culture

I would like to present this chapter in keeping with what Yaghejian (2002) describes as introducing a particular voice while demonstrating an understanding of the concept of Participatory Action Research as I, as a member of the community, have defined it. Yaghejian (2002) writes that

> the act of writing home refers to the process of creating home that becomes possible through literacy, since reading and writing involve not only a product, but a process of reproduction. As such, *Writing Home* illustrates the need to revitalize the heritage language through creative practice. (p. 4)

Unfortunately, as I still have trouble finding a Kabyle philosophical or spiritual framework, I will employ a Western academic framework for now. To begin, I write home using narratives about moments that my family, my community, or I myself have been engaged in. Shirinian (2000) describes beginning as follows: "Beginning implies both return and repetition and not a linear development to some ultimate moment...Beginning a quest, departure from what is known, sets up a connection between continuance and continuity" (p. 145). In the same vein, O'Reilly-Scanlon (2000) writes that

> the real value of this kind of reflection lies not within the past, but within the realm of possibilities and potential to change the Future. Acknowledging that the past is a prologue to our future we need to examine and question how our understanding of what we have learned about our pasts and how the formation of our selves can effect change to make us better people and teachers. (p. 206)

Before addressing CPAR, Kabyle educational experiences, and research possibilities, let me provide a brief description of the historical and political context of my community, the Kabyle, in Algeria.

The Historical and Political Relationship between the Kabyle and Colonizing Groups and the State in Algeria

I am Kabyle. I was born in and grew up in France. I used to live in a city whose economy was based on the extraction of minerals and on metal production. My father was brought there during the colonization and worked to rebuild France after the Second World War, like most of the men of my village in Algeria. The process of importing labor to France has created a Kabyle diaspora, which became the context of my upbringing. In this Kabyle diaspora, I heard many stories about my land. My family in France and Algeria have talked in great depth about my ancestors; ancestors who were seen as regarded highly within the community and were said to have tremendous healing powers.

Coming from the last village in the range of the Djur Djura Mountains, I am a Marabout, a man of very ancient heritage, linked officially today to the bloodline of the prophet Mahomet; a heritage questioned in many ways today. Each village had a man as a leader, an educated and spiritual man considered by the community to be a saint. The stories say a lot about my great great grandfather. It is said that one day, the village was having a meeting to decide about splitting up. My ancestor did not support the project of splitting up but the conversation turned against him. Upset, he exclaimed, "Take your share and I will take mine. Everyone who wants to stay with me should follow me." When he left, the mountain followed him. The story explains how the village then decided not to split up and it is now one of the reasons given to explain the actual shape of that particular mountain.

I find this story very interesting because it shows the image of a traditional political organization, a dynamic between Kabyle people in the mountain, and it demonstrates a system of belief that brings the community together as one with the land.

The saints and founders of those villages are considered to be descendants of the prophet Mahomet. They all have a sacred place where people go to ask for healing and blessings. Even if the grave of the ancestor is known and protected by the residents of the village and is considered an important Islamic place by the people—a place for collective contemplation—it is denounced by the Algerian government. Also, the dominant practice of Islam in Algeria considers the villagers' practices to be part of a non-Islamic cult. This condemnation compels people to attend the mosques in the city where the only language promoted is Arabic.

However, regardless of such pressures, in the mountains there is a strong tradition of shamanism that proves the existence of another system of belief that is closely related to the original *Amazigh* culture. Camille Lacoste-Dujardin (1984) writes that "Tamazight [a Berber language] supports a very big and rich oral literature in prose—myths, ante Islamic tales, legends, nice stories—and poems (love poetics, songs of work, of celebrations, songs for babies, etc.), and the long local persistence of the oral has been kept alive until now" (p. 65).

To begin any kind of educational project in a mountain community in Kabyle, it is really important to consider the presence of animism and Islam, two strong and deeply rooted spiritual and/or religious traditions, which shape the social and political situation in Kabyle. However, looking into the political history of the mountain can also be of great help in this regard. Also, the works of Nedjma Abdelfettah Lalmi (2004) and Lacoste-Dujardin (1984, 1999) can help us understand political foundations of Kabyle society today.

Nedjma (2004) talks about a "community of the mountainous relief, about the use of Berber language, about sedentary life style and about the existence of a municipal tradition as well as about a usual system of rights distinct from the Muslim system of right" (p. 511). She also supports the idea that a Kabyle state existed and had Bejaia as a capital long before French colonization. She quotes Emile-Felix Gautier who talks about a "Kabyle kingdom... one of the faces of the Senhadla Kingdom, this other one being the Ifriqya (actual Tunisia)" (p. 514). For Gautier, the quality of this relationship proves that this state was not a stranger to the tribes and that "Bejaia was their own Capital city" (as cited in Nedjma, 2004, p. 514). Nedjma finds artistic evidence of this within the art of Guergour: "This link to the world and to the city is being confirmed by the art of Guergour, art of synthesis between geometrical berber forms and the round and floral motifs that surprised Lucien Golvin (1955) in his study of the carpet of the area" (p. 524).

Lacoste-Dujardin (1984) supports the idea of an opposition between Kabyle and Arabs by saying that "to understand, we need to go back to 1847 when, preceding the conquest of Kabyle, Bugeaud launched some kind of ethno-political Kabyle from a representation opposing 'Sedentary Berbers' and 'Arab Nomads.' We need to add that Kabyles resisted the French army for a long period, without joining Abd el Kader" (Lacoste-Dujardin, 1984, pp. 257–277). Nedjma (2004) explains that the word Kabyle refers to an involved relationship between Kabyles and Arabs. She says,

> In the same time that Maghrebian cities were developing and becoming Arab, their backcountry was designed as territories of the "qbail", which means tribes. It is why we can find "qbails" around Bejaia, Jijel, Kolel, Tlemencen, Kolea, Cherchell, etc. According to G. Yver (1927), it is in the Qirtâs (Ibn Abi Zar', 1860), that the word appeared for the first time as an ethnic reference. Tribalism and berberism, or berber fact or tribal fact, become then pretty much synonymic and in the actual Kabylie, *qbaïli* (man of the tribe) becomes an ethnonym with which the indigenous are going to identify themselves. The universal oppositional ratio between cities and rural land, is going to be taken more like an ethnic opposition between Kabyles/Arabs because of the fact of a progressive arabisation of the cities. (p. 513)

Expressing the agenda of the two main political pressure systems that the *Amazigh* colonial and/or postcolonial society has had to support in Algeria, the Islamic/Arabic one and the French one, Nedjma gives us a great deal of information about the political and cultural situation of Algeria today. She digs in to the Muslim foundations in the region during the middle ages and looks at the relations between Arabs and Kabyles since the early twelfth century to explain the cultural situation as well as the political heritage of the *Amazigh*. She comes to the conclusion that

> It appears to us that the split with the central state is due to... the loss of Bejaia in 1510 to the Spanish—and it contributes to explaining the development of a phenomena that, as we know it, has only been studied in synchronical studies of anthropologists and contemporanists, known as the "village republics" of Kabyles and their assemblies. (2004, p. 516)

Talking about the war of independence in Algeria, Lacoste-Dujardin (1999) refers to Messali Hadj, the leader of the independence movement in Algeria in the early 1900s. Messali Hadj led a party that changed names a couple of times because of French interference but it was known for a long while as the MNA or Algerian National Movement. He was pushing for a Muslim, socialist, and democratic state. This movement was supported for a long period by people like Hocine Ait Ahmed, leader of the actual FFS or Front of the Socialist Forces and a great number of *Imazighen* (free men). In fact, in 1936, according to Lacoste-Dujardin (1999), 70 percent of his party consisted of Kabyles, a proportion that was equivalent to Kabyle emigration to France (2001).

Talking about education in our communities, Nedjma points to the existence of a strong tradition of educated people in Kabyle because of what she considers the Muslim precolonial society that in turn also helped explain why there was a positive response to French schooling initiatives. Kabyle people have always had a high regard for education, as suggested by Nedjma and Lacoste-Dujardin. Through the ages we have demonstrated a great ability to adapt, be creative, and remain great producers of culture and knowledge. This process needs to be supported and protected today. CPAR opens such possibilities in the current Kabyle political, social, and cultural context as it is research that is rooted in the Kabyle values of sharing and collectivity that are so important in our villages. As Lacoste-Dujardin (1970) puts it,

> The egalitarianism of the Kabyle village residents shown through the oral literature reveals in every occasion a real obsession for the appropriation of personal power against which the village community's fight has been constant. The most celebrated heroes in the communities are before everything very careful about restoring the community's order, fraternal and egalitarian. (p. 534)

When it comes to CPAR and the prospects for Kabyle education, the possibilities need to be informed by the current situation in the classroom. Schooling only in Arabic is maintained through centralized control by a state that actively suppresses all local possibilities and compels assimilation in to the dominant Arab educational system. For example, while school budgets are apparently at the discretion of the school principal, principals are under surveillance by the local government. The pedagogical implications of this politics of cultural control were the primary motivation for the following and preliminary CPAR that I initiated in the spring/summer of 2008.

Education: Kabyle Is Taught as a Foreign Language

I met the principal of the elementary school while traveling in the mountains to help my uncle with his biweekly selling journey. The principal is from the school that children from my village attend. My uncle is a merchant and one of my great pleasures is to accompany him to the mountains and meet the people and listen to their stories. Although many others have tried to open their own businesses to compete with him, they have all failed. Meanwhile, my uncle has not even had to change his prices. This is why he is known and respected. People also know of his involvement in the village, and they owe him a lot because he stayed with them even when times were ridden with conflict. They trust him and this is what is valued in our village, more than any thing else.

My uncle introduced me to the principal of the local elementary school when we were traveling together in the mountains. The principal gave me an appointment to meet with him at the school. On the day of the meeting, he arrived a bit late and could not open the door because he did not have the keys of the school with him. He asked for the school guard but none of the people sitting in front of the school had seen the man. The school is located on one of the rare cross roads in the mountain and near the unique and easily accessible fountain in the area. In the summer, this place is really significant because of severe water shortages in all the homes. It was actually after a community meeting that the people of the village decided to build this installation. I remember when I was young that every time somebody was thirsty in the village, they would send a child to go knock on doors and get water.

That day the principal was in a very delicate situation—he had somebody he promised to show the school to, a school, no keys, and no guard. He called a child in the street by his name and asked him to run to the guard's house to get the key. The child was really happy and ran to do as he was asked, clearly demonstrating his affection for the principal. I was happy to have this chance in the middle of the summer to visit the school with the principal. I was already playing with the kids every day. The school is a very basic building with really simple and minimal furniture. The official texts are all written in Arabic and the classes are run in the same language. It is a problem when alphabetism (in Arabic) is given so much importance in an area where the only language spoken by families is Kabyle (an oral tradition). French is being used less frequently and primarily on occasions and for matters outside of the local Kabyle culture.

During my school visit, we were talking about the religious posters in the local building where my ancestor is buried. I shared what I thought about it and elaborated on how and why fundamentalist Islamism was making its ways in to our community and how this was replacing our beliefs and entering our systems of representation. Islamists have constructed a building that looks like a mosque around the gravesite so that people can see it from afar and for the few who make it inside, they have even put up posters of the pilgrimage to Mecca.

In Kabyle, there is a complimentary tension at play on both the Berber language cultural and fronts. Kabyles have expressed their need for official recognition of their identity but the response has been a partial one and one that is primarily focused on diminishing sociocultural-linguistic tensions that have been created in Kabyle by state control over Kabyle education. For instance, children from Kabyle have to wait until junior high school to get a chance to learn their language in school. Kabyle is then offered as a supplementary subject for four years. It is taught by teachers who often might know Kabyle but are not necessarily pedagogues who know how to teach the language in school settings. There is no teaching of Kabyle identity and in fact, the educational process continues to reproduce a social pressure to conform to Arab ways while actively working to fragment Berber identity. As Bernstein (1990) observes in this regard,

> I have argued that the assumptions of invisible pedagogies as they inform spatial, temporal and control grids are less likely to be met in class or ethnically disadvantaged groups, and as a consequence the child is likely to misread the cultural and cognitive significance of such a classroom practice, and the teacher is likely to misread the cultural and cognitive significance of the child. (p. 84)

Or, as Wertsch (1991) would likely conclude, "Given the difficulty, it is perhaps surprising that students are ever able to recognize, let alone master, the formal instructional speech genres grounded in decontextualized word meaning" (p. 135). More than this, teaching Tamazight, which is in the family of Berber languages, in this manner in the classroom potentially creates the feeling that Tamazight is an option (out of other possibilities) and a burden for the development of children who could, in the end, be successful without really having to study it. Ironically, while it is considered a victory by some Berbers that the local language is even being taught at all in the formal schools, this approach to teaching actually retards the moment when Tamazight might become the official language in our schools. At the same time it increases the social exclusion of Kabyles from the educational process and the dominant Arab society of the Algerian elite. So, as the principal of the school was saying to me, "in Kabyle, Kabyle is taught as a foreign language."

The pedagogical materials remain purely linear and oriented toward basic knowledge acquisition. Freire and Macedo (1987) tell us that

> Mechanically memorizing the description of an object does not constitute knowing the object. That is why reading a text as pure description of an object (like a syntactical rule), and undertaking to memorize the description is neither real nor does it result in knowledge of the object to which the text refers. (p. 33)

This pedagogical approach, received from a colonial era, is inappropriate as it objectifies the *Amazigh* identity while silencing the educative and cultural claims of the local families who seek social change in the form of official recognition for Berber language and culture. The reality, however, is a system that reinforces the notion of outside control and authority over the child's home and community culture. The official curriculum shuns *Amazigh* culture by encouraging an educational process that works to cause the culture to fade away in the mind of the child, while also teaching them knowledge perspectives that are inappropriate in relation to the *Amazigh* system of knowledge production. As Wertsch (1991) puts it,

> The negotiation of referential perspective in this case required more change on the part of the children than on the part of the teacher, something that correlates with the difference in power between the two interlocutors. The movement over the course of this segment of interaction has been one of children giving way to the teacher: they have capitulated to her use of scientific concepts as an appropriate grounding for describing objects. (p. 117)

This year, Kabyles wanted to commemorate an incident that took place in the spring of 2001 when police shot a young Berber during an agitation denouncing abuses that the police were committing in Kabyle. The message in Algeria is always consistent: that power is in the hands of the government and only their politics count. Any other social movement or way will be treated as a threat and will be repressed through brute force. It is a form of rule and way of life that is opposed to the *Amazigh* language and culture and finds its expression in the actual education system.

Conversations around a Scholarship: CPAR and an Initiative for Berber Education and Culture in Kabyle

Research, whether it is participatory or not, is always difficult in a context like the Berber one. The political context of war, manipulation, betrayal, and torture that the mountains have endured is now part of our culture as are suspicion and reserve to any form of intrusion (e.g., research) as boundaries are drawn but always with a sense that there is still a great will from Kabyles to make our situation improve. As Shirinian (2000) expresses,

> The concept of a frontier/boundary is a rich ambiguous semantic field that includes binary oppositional pairs such as separation and belonging, opening and closing; it is a barrier and at the same time, a place where communication and exchange occur. It is a site of dissociation and association, separation and articulation. The frontier/boundary is a filter by which elements are blocked or allowed to pass. (pp. 57–58)

I have had to live with these boundaries in different ways. Recognizing that Kabyle culture, far from being a handicap, is definitely the foundation for a better quality of life for the Berber people, I decided to address cultural boundaries by proposing the idea of a scholarship to support a child in high school. The emphasis of the scholarship was to be on Berber cultural engagement (like going back to the village, participating in traditional moments), academic results, and general participation in school life. The message of the scholarship was to emphasize that it was possible to embrace Berber culture and succeed and that the two were possible together—a message that contradicted the state's assimilationist approach to Berber education and culture.

Since the scholarship was supposed to be given to a child from my village, I needed to go through the customary process, meaning that any individual initiative on my part would have been a mistake. My father reacted in an exaggerated way to my initiative and decided that it was a waste of money. He told me that people did not even have drinkable water and that such an education was a luxury. In my house, for instance, in the middle of the summer we only have access to water every three days.

Two events helped me get the scholarship proposition accepted. First, I chose to endow the scholarship in my father's name thereby showing respect for my lineage. Second, I had a dream that I shared with a member of the village. In the dream, I had to face a community meeting where I was supposed to explain my strained relations with my father. In my dream, I could see people with familiar faces but also other people, the shadows of people that had already gone by. They were listening to the debate, a debate in which I was silenced by the rhetorical tactics of my father and my uncle. I remember feeling extremely upset and sad for all of us. I could see the roots of our pain and suffering in this kind of behavior, behavior that could, may be, be explained by the fact that they never had a chance to heal from the traumatic experiences of the past. The council decided to chase me out of the village. So I went to the entrance of the traditional trail that was going from the top of the mountain to the valley. I was about to leave but then decided to fight because I felt that they were being manipulated and that I needed to stay with them and with my ancestors. As soon as I made that decision, my ancestor came to me and I saw myself standing beside his grave. He told me clearly to stay with him and that he would take care of me and teach me.

Not knowing really what to make of that dream, I shared it with somebody else who then repeated it to another person and soon the dream went all over the family like fire running on gunpowder. My ancestors had spoken clearly and expressed clearly that my place was with him and this helped my father to accept me back. Subsequently, I have been able to pursue the scholarship project.

In the same way that I was sharing my scholarship idea with my father, I told him about writing this chapter on CPAR. Again, he reacted strongly and accused me of spying on our people. He told me that he had already gone through a lot during the war, and he did not want his family to live the same story. Doing research in and writing about Berber communities can have many consequences for us.

When the student who eventually received the scholarship visited me at home, my father asked him if he would be capable of resisting the torture and by this he was referring to state treatment toward him and the Berber during the war of independence—he was anticipating difficulty for anyone who might chose to resist the designs of the state, including through acceptance of a scholarship that seeks to assert Berber ways in education. I asked the student if he went to the village often and he said he did not but said that he attended the mosque regularly. Giving up the official and moral tone of our conversation, I spoke to him about family and friends. I told him how thankful I was for the time I spent with him, especially after a four-year absence from my village—four years because I did not have the army papers proving accomplishment of military duty, and without those, I could not cross the border without being taken by force and sent to the streets to fight terrorism.

The student was a young boy, perhaps eleven years of age. His father asked him to take me to our ancestor's grave. While walking up the mountain, we passed by several villages where people would always ask who I was and thereafter, wish me a warm welcome back. It is when the boy finally felt comfortable with me, that he put his hand in his pocket and took out a key chain with the symbol of the *Amazigh* people on it and offered it to me. It is because we are all part of the same family that he accepted my support and shared with me what he and his family find so important—our traditions, our land, and our *Amazigh* identity.

This community attachment and this sense of belonging that brings us together is always undermined, and it is considered by outsiders as naive and retrograde. Our traditions, however, do not accept individualism and the success of one is seen to bring the happiness of all. Our methods to keep our society together generate and protect a knowledge that is available to all who can listen like *Amazigh* people. Connected to our land and history, it is not a knowledge that *Amazigh* want to universalize but one that can and does participate in the safe keeping of a universal balance.

Unfortunately, the government does not see it that way and it develops a strategy to counter *Amazigh*, a strategy that can be described through Bower's (1998) analysis when he observes that "the influence of social, self sufficient communities would have to be undermined in order for the individual to become 'educated'" (p. 7). By presenting Arabic civilization as the first and only civilization and as scientific, modern, and developed in opposition to a so-called undeveloped and primitive Amazigh society, the government is undermining a rich culture that can bring much more to the world than it is permitted to.

Amazigh life and ways contradict the corrupt necessities of the Industrial Revolution and related emphasis on a particular type of individualized education. In Bower's (1998) words,

> The Industrial Revolution required a radically different form of individualism, one that took-for-granted the following assumptions: that education leads to the individual becoming an autonomous, rational thinker capable of judging the merit of community traditions; that progress is linear and that the high status knowledge learned in the classroom represents the more evolved stage of cultural development. (p. 7)

Located outside the official and abusively legitimated colonial system of knowledge production, *Imazighen* survived the industrial revolution by keeping their language, their link to the land, their families, and their traditions. The government that argues "that de-contextualized print-based knowledge and forms of communication are more reliable and culturally advanced than what is learned in face-to face relationships" (Bower, 1998, p. 7). These knowledge forms are trying to break our social system by teaching *Amazigh* culture and language as a subject that is scientifically debated in the formal classroom. This pedagogical approach is destroying our families and cutting the links to our heritage. This same government politic imposes its ethnocentric ideas pretending,

> that the veracity of ideas and values should be determined in an open, competitive environment; that the narratives, processes of inquiry, and technological innovations learned in classrooms should be based on an anthropocentric view of the world that the epistemology of science and the systems of expert knowledge provide the more reliable forms of knowledge for rationally managing the internal and external world—and that the resulting systems of commoditization should be globalized. (Bower, 1998, p. 7)

Bower's analysis of the words of industrialization illustrates some of the industrial-cultural traps that we, the *Amazigh*, are trying to avoid. The changes that *Amazigh* society might need are changes that will bring the community back together with an education that will help overcome issues germinating from the colonial trauma while simultaneously inscribing *Amazigh* identity in the world. This is the defining project for CPAR with Kabyle—a long-term community project of a people's research, action, and education that helps protect our human rights and lifts us out of economic poverty in a culturally and ecological manner that is consistent with our ways.

Concluding Reflections on CPAR and Amazigh Culture and Education

In the village where we/I are trying to support a regenerating cultural dynamic to address the colonization of *Amazigh* knowledge, there will continue to be those who will try to obstruct such changes from taking place. People happy with the status quo accept a government that will try to break the links between culture and history to replace it with a foreign culture and religious people who will try to make us abandon our spiritual roots for a politicoreligious organization.

However, the energy coming from a strong *Amazigh* tradition is still there, and it is a healing energy that can fuel a significant process of social change. Returning to Algeria and my village raised many questions regarding how I could participate in and support my culture, my traditions, and how I could help to develop an education system based on our ways. CPAR embedded in the historical, political, social, and spiritual location of my people can provide a small way forward. A brief encounter with an elementary school exposes the suppressive pedagogical dynamic at work in governmental schools in Kabyle. The idea of a scholarship for a high school student of the village leaves me with the feeling that the road to cultural recovery is long but that the will is there, hidden in the hearts of the children and the mountains.

The connection between land and people is common to many indigenous nations. My participation in the meetings of my village, my life within the Midiwin society in North America and the teaching lodges, as well as my participation in the *Marae* (sacred lodge) in the Maori culture of New Zealand have revealed some very interesting cultural possibilities. *Tajmaet* and/or *Zaouia* (gathering times and places), the Midiwin lodge, and the *Marae* are all places of spiritual education as well as transmission and creation of knowledge. Designed in similar ways, they demonstrate the interconnectedness of past, present, and future, the transmission of knowledge, and the links with the ancestors. They represent a cultural-ecological perspective and epistemology that is taught in ceremonies but expressed in everyday life. Deeply rooted living cultures that are strongly connected to their various contexts, their similarities also reveal an indigenous epistemology that is being choked (colonized) beyond recognition but persists given the resilience of those that continue the struggle—a struggle that has a long way to go yet in the protection of our ways.

References

Bernstein, B. (1990). *The structuring pedagogy of the discourse: Volume IV. Class, codes and control.* London: Routledge.

Bower, C. A. (1998). An open letter on the double binds in educational reform. *Wild Duck Review,* 4(2/Spring/Summer), 7.

Freire, P., & Macedo, D. (1987). *Literacy: Regarding the word and the world.* South Hadley, MA: Bergin & Garvey.

Lacoste-Dujardin, C. (1970). *Le conte kabyle. Étude ethnologique* [The Kabyle tale: Ethnological study]. Paris, France: François Maspero.

———. (1984). *Genèse et évolution d'une représentation géopolitique: L'imagerie kabyle à travers la production bibliographique de 1840 à 1891.* [Genesis and evolution of a geopolitical representation: Kabyle imagery through bibliographical production from 1840 to 1891]. In *Connaissances du Maghreb,* CNRS, Paris (pp. 257–277).

———. (1999). *Une intelligentsia kabyle en France: Des artisans d'un "pont transméditerranéen"* [A Kabyle intelligentsia in France: Craftsmen of a transmediterranean bridge]. *Hérodote,* 94(3), 37–45.

———. (2001). *Géographie culturelle et géopolitique en Kabylie. La révolte de la jeunesse kabyle pour une Algérie démocratique.* [Cultural geography and geopolitics in Kabyle: The revolt of Kabyle youth for a democratic Algeria]. *Hérodote,* 103(4), 57–91.

Nedjma Abdelfettah Lalmi. (2004). Du mythe de l'isolat Kabyle [From the myth of Kabyle isolation]. *Cahiers d'Études Africaines, 175,* 507–531. Retrieved from http://etudesafricaines.revues.org/document4710.html

O'Reilly Scanlon, K. (2000). *She's still on my mind: Teachers' memory-work and self-study.* Unpublished Doctoral Dissertation, Department of Integrated Studies in Education, McGill University, Montreal, Canada.

Péguy, C. (1910/1993). *Notre jeunesse.* France: Éditions Gallimard.

Shirinian, I. (2000). *Writing and memory: The search for home in Armenian diaspora literature as cultural practice.* Kingston, ON: Blue Heron Press.

Wertsch, J. V. (1991). *Voices of the mind: A sociocultural approach to mediated action.* Cambridge, MA: Harvard University Press.

Yaghejian, A. (2002). *From both sides a border, writing home: The autoethnography of an Armenian-Canadian.* Unpublished Master's Thesis, Department of Integrated Studies in Education, McGill University, Montreal, Canada.

CHAPTER FIFTEEN

Notes and Queries for an Activist Street Anthropology: Street Resistance, Gringopolítica, and the Quest for Subaltern Visions in Salvador da Bahia, Brazil

SAMUEL VEISSIÈRE

Preamble

Salvador, Bahia, July 2008

"No need to go to the police.... I could take care of him for you, you know?" declaims Emilio calmly, motioning his chin toward Neguinho's fleeting figure, who has now hurled his makeshift glue inhaler against the sidewalk and stormed away in anger, cursing in loud shrieks.

Emilio, who is well known and respected in the neighbourhood, spends his days stationed on a street corner where he is employed as the plainclothes security guard of an upmarket café popular with gringos and the local bourgeoisie. When the café closes at night, without going home or changing the red denim shirt he has now been wearing for two weeks, Emilio takes his post three hundred yards further, next to a British pub by the seafront. There, until two, sometimes three in the morning, he makes sure the gringos and wealthy Brazilians who get dropped off from air-conditioned vehicles can cross the sidewalk safely and enter the air-conditioned haven of the British pub with minimal exposure to the street.

Neguinho, whom I have known for more than two years as an informant, and as I would have once described him, a friend, used to sell bracelets and necklaces with his older brothers, navigating the bar-front terraces until the early morning hours. There, with his angelic face and skilfully performed cute-child looks, he would consistently outdo his brothers in sales, and be treated to food and drinks by teary-eyed gringos. Those days are gone now. Neguinho has traded his designer cap and clothes for ragged swimming trunks, bare feet, and a T-shirt sporting the name of a political campaigner, long since defeated in exurban local elections. Of a child, Neguinho has retained the stature, but in the bulging veins of his scrawny neck and his blood-shot eyes, there is nothing angelic anymore. Neguinho, as street language has it, "*está no crack*" now; he has fallen "into crack", which also means he's a goner.

In two separate events in the past week, he robbed two of my friends; one of them a researcher, both of them gringos, both of them having travelled to Salvador to see me. Tonight, my friend Miles (the ethnographer) has managed to corner Neguinho, demanding explanations and retribution. Two other street kids have come to Neguinho's rescue, and Emilio has now been summoned.

We know where Emilio stands on this issue: "It is because of little fuckers like him," he had claimed in a previous conversation, "that this neighbourhood has a bad name. Soon, gringos won't want to come around here at all. Little fuckers like him are bad for business, bad for gringos, and bad for hard-working people." Now, despite his composed look, Emilio seems furious, and is offering to take the matter in his hands:

"…the police will just give him a good beating, but I can take care of him, you understand…?" As he speaks, he throws a knowing glance toward his crotch, above which dangles his untucked red shirt, and his hand reaches for an imaginary spot above his hip. Watching his finger pull an imaginary trigger, I contemplate the weight of his words:

"…I mean *really* take care of him…."

How did I get here? How did I go from wanting to research, understand, and transform a particular sociopolitical context—in this case the livelihoods of street peoples in urban Brazil—and later, as a result of my very presence in this context, participate in a chain of events in which the next logical step would appear to have led to the murder of a child?

Introduction: *Conscientização* and the Quest for Subaltern Visions

What are the implications (ethical, political, physical even) when researchers and activists situated in geographic, economic, political, cultural, and epistemic positions of power enter the field in the Global South with the desire to facilitate participatory and emancipatory forms of action-research among marginalized populations? What are the implications and complications of entering the field with a priori, primarily theoretical, and often culturally displaced visions of social change? What are the implications for the researcher, the informants, and participants, and the broader context in which the research takes place?

Stemming from my fieldwork with street kids in Salvador da Bahia, this chapter presents reflective insights on the possibility and politics of engaged, activist, and participatory social science concerned with grass-roots social transformation. Focusing on the complex politics of interaction between foreigners (researchers, tourists, and activists), street populations (sex-workers, hustlers, street kids), and other local and transnational actors (NGOs, social movements, police, business owners, etc.) in Salvador, I examine the different visions of change that are imagined and desired from above and transposed onto activists' agendas aimed at empowering, transforming, or cleaning up the street. By street I refer to this partly discursive, partly spatial, partly transnational, and deeply contested site of struggle, both for the people who live, survive, and make a living in it, and those who conceptualize it as a place where social change must originate.

As a gringo ethnographer ambiguously positioned within this transnational sphere of interaction—an interactional and political space that, inspired by Diane Nelson's (1999) discussion of her "gringa positioning" as a would-be activist anthropologist in Guatemala, I term *gringopolítica*—I also pay particular attention to subaltern social critiques voiced by street people themselves; I keep an especially attuned ear for spontaneous forms of conscientizations, visions for change, and acts of resistance that emerge from the street, and do not echo the hegemonic discourse of NGOs, social movements, gringo activism, and other external articulations.

By focusing on seemingly isolated, disorganized, and highly informal acts of resistance from the street and contrasting them with other visions of change, this chapter does not propose any model or report on any successful case study for participatory research. It remains fundamentally concerned, however, with a concept that is central to the project of critical, applied, and emancipatory scholarship: conscientization.

The praxis of conscientization, stemming from Paulo Freire's discussion of *conscientização* in his *Pedagogy of the Oppressed* (Freire, 2000 [1970]) can be defined as "the process of becoming aware of the structural, political, and cultural constraints that prevent a group or individual from exercising autonomy or participating in a democratic society, and the subsequent practice of working toward emancipation" (my definition). Most scholars working in the Freirian tradition of Critical Pedagogy, in turn, have focused on ways to build more inclusive societies by facilitating the conscientization of oppressed people in both formal and informal, or popular, spaces of education (such as schools and universities, or NGOs and social movements; see Abdi & Kapoor, 2008; McLaren, 1997).

In my own work as an anthropologist and would-be-activist, I am involved first and foremost in a search for spontaneous forms of *conscientização* that emerge in contexts of struggle. I have also become interested in—or concerned with—the politics and dangers of imagining, articulating, and facilitating *conscientização* on other people's behalf. Finally, I am concerned with ways to reconcile the insights of subaltern people themselves with the level of structural power that is available to academic researchers, and with the possibility of nonintrusive and nonhegemonic emancipatory research.

In this chapter, I argue that any critical and nonhegemonic participatory agenda must be articulated around insights that are both anthropologically informed and grounded in strategically realist visions gathered from in-depth, informal, and long-term ethnographic fieldwork and connections with the population one hopes to empower. I contend that a researcher intent on devising participatory projects that will benefit marginalized people should have earned quasi-insider's perspectives through long-term ethnography or personal relationships, or refrain from getting involved altogether.

This reflective piece, thus, turns the question of participatory praxis on its head, and instead of advocating the transcendence of theory through activist practice, presents the story of journey away from *disengaged practice* and invites researchers to consider the participatory and emancipatory potential of less immediately ambitious, but deeply qualitative forms of ethnographic research I term *engaged theory*. Thus, I invite researchers and activists to reflect on their own positioning and question their own motives and visions for participating in grassroots social change, and specifically, I urge researcher-activists to pay particular attention to the role, visions, and levels of critical consciousness and authority they attribute to the very people they hope to "empower."

I do so by returning to the pertinent question posed by anthropologist James Ferguson (1994) in his study of bureaucratic power in Lesotho in which he famously speculated on the role of intellectuals in subaltern struggles. Reflecting on whether the presence of intellectuals would be adequate, or even welcome in the context of locally responsible "development," he proposed that the most important question we should ask about development was not "what is to be done?" but simply, "by whom?" While I laud the pertinence of Ferguson's important question, I am also, given the urgency of many of the issues we have to confront as engaged scholars in postcolonial times, tempted to argue in favor of a strategic move beyond the paralysis of postmodernist obsessions with power and co-optation, and therefore continue to argue that intellectuals who become implicated in postcolonial contexts are not so much entitled to, but morally obliged to contribute to the social struggle of the marginalized people they study.

Indeed, I have adopted a similarly pragmatic position in relation to the irresolvable questions of objectivity and power made naggingly apparent in the postcolonial critiques of ethnographic authority that have compelled many younger ethnographers to refrain from writing on other people's behalf altogether (notably beginning with Clifford & Marcus, 1986). Doing fieldwork in dangerous places among peoples with no political voice, thus, I have learned to settle for a strategically realist and engaged form of ethnography in which I do, without qualms, borrow other people's voices to tell stories that need to be heard. Nancy Scheper-Hughes' (1992) similarly strategic and most eloquent call for "good-enough ethnography in perilous times," voiced nearly two decades ago, is worth retelling at length:

> I grow weary of these postmodernist critiques, and given the perilous times in which we and our subjects live, I am inclined toward a compromise that calls for a "good enough ethnography". The anthropologist is an instrument of cultural translation that is necessarily flawed and biased. We cannot rid ourselves of the cultural self we bring with us into the field anymore than we can disown the eyes, ears and skin through which we take in our intuitive perceptions about the new and strange world we have entered. Nonetheless, like every other master artisan (and I dare say that at best, we are this), we struggle to do the best we can with the limited resources we have at hand—our ability to observe carefully, empathically, and compassionately. (p. 28)

Like Nancy Scheper-Hughes on ethnography, I am also inclined toward a compromise that brushes paralyzing postmodernist critiques aside, and calls for "good enough activism for perilous time." Yet, I have also learned through the experience of fieldwork that caution and restraint to get involved are at least as important as strategic pragmatism. Thus I now engage myself in activist research without dismissing Ferguson's question, but rather by turning it around and instead of "who is to get involved?" I ask, "according to whose visions should social change be planned?"

Gringopolítica as Hegemony: From Disengaged Practice to Engaged Theory

Let us examine the question of how participatory projects inspired by Critical Pedagogy might accommodate different visions of change. To help us situate this project closer to the context of my research, I borrow the words of Fátima, a character in Tobias Hecht's (2006) *After Life,* the ethnographic novel set in the streets of Recife in Northeast Brazil that provided some of the inspiration for my own work. Fátima's words, I find, capture the essence of the role of researcher-educators preoccupied with a critical exploration and collective transformation of reality, and the philosophy of many of the activists I encountered in the field:

> If you go into a community and ask people what they need, they will tell you a soccer field. That's what they feel they need. But if you engage them in a process of reflection about their lives, they will see

that what they really need is to vaccinate their children, cover up the open sewerage lines, reduce the incidence of domestic violence. As an educator, *um educador*, someone working in the tradition of Paulo Freire's Pedagogy of the Oppressed, one's job is to help people distinguish between one sort of need and another. (Hecht, 2006, p. 174)

Such a philosophy—that of the critical *educador*—has always animated the way I teach and do research. Yet after experiences of researching, teaching, and wanting to bring about change in many different postcolonial contexts, I came to question the extent to which such methods truly allow for collective or participatory transformations of reality, and I became increasingly troubled by the dangerous amounts of pretension inherent in the position of the researcher-*educador*. Gradually, I came to recognize the inherent condescension, and even danger of such a pedagogical philosophy.

To be sure, if one's role is to "guide" others beyond deceptive or oppressive aspects of social reality, it follows that one must claim some form of superior knowledge of social reality. This rather arrogant posture, as James Scott (1990) reminds us, is necessarily grounded in presuppositions of "false consciousness" that deny social actors all authority over their own realities and priorities (p. 70). Spending time in Salvador among educators, NGO activists and other would be do-gooders who invoked an allegiance to the Freirian tradition in their street practice, I found the concept of false needs to be a recurring theme. Many activists, as I came to understand, conceptualized their mission as a quest to guide the children beyond the false needs of the street.

The true path toward liberation, as I discovered, largely depended on the ideological position of the activists, NGOs, or social movements, and thus, could entail such varying propositions as mobilizing in the name of the proletariat (Marxists); mobilizing along racial lines to rediscover and perform ones' African roots in a variety of politicized or kinestheticized ways (revivalists); foregoing one's idolatrous and demoniac African practices for a personal relationship with Jesus (evangelicals); finding a connection with the suffering of the poor and path toward liberation through Jesus (liberation theologians); finding wealth and happiness through Jesus (evangelicals again); or becoming micro-entrepreneurs, going to school, learning English, working in the tourism industry, or, simply, staying off the street.

Confronted with these conflicting visions and fantasies of social change imposed on the street from external actors, I also became increasingly troubled by two questions. The first was one of positioning: How did I, or could I, as a gringo academic, fit into this quest for social change? The second had to do with the general obsession with the street: In a context of such vast social injustice, why were so many activists focusing on the narrow question of children being in the street, and the need to take them off the street? Through a series of what can only be called epiphanies,

I began to disengage myself altogether from established organizations and social movements, and shy away from the idea of working toward any form of mobilization based on my own presuppositions.

The first epiphany occurred under the form of a rejection in the early stages of my research. It stemmed from a conversation with the director of a prominent NGO, explicitly Freirian and African-revivalist in its mandate, with which I had hoped to be associated as a street educator. "You should know," the coordinator had told me before inviting me in his office, "that we no longer accept volunteers, and especially not foreign volunteers." Filled with shame, I had remembered Tobias Hecht's similar experience with a social worker in Recife who had told him that "like many foreign visitors, [he—Hecht—had] come to study [Brazilian] misery" (Hecht, 1998, p. 17). Before I could offer him a reply, the coordinator continued, "we had too many problems of a *cultural* order... too many people who wanted to *show us the way*." Later in his office, as he went further in voicing his frustration at gringo obsessions with street kids, I chose to remain silent. Why defend myself, it seemed to me, if I could only agree with the coordinator's concerns? Here I was feeling more than ever like another street-kid tourist in this global mess Tobias Hecht (1998) called the "street children industry."

I had, before entering the field, some awareness of the problematic politics of the gringo obsession with the street. This had been based both on readings of anthropological literature and on my previous informal interaction with street kids. I understood that, in a global context where, as Arjun Appadurai (1996) had famously commented, modernity was at best an unevenly experienced phenomenon, children in the slums and streets of the Global South had come to represent the quintessential residual category of globalization (Scheper-Hughes & Sargent, 1998). As such, and unlike the plight of many others among (post)modernity's discontents, I knew that the idea of street children engineered by sensational media accounts and other poster-child rhetorical devices had created an eerily unanimous sense of discomfort among citizens of the North, and that as a result, street children had become the object—if never quite the subjects—of countless charity campaigns and many other education, research, rescue, and salvation endeavors.

I knew that a similar idea of street children had also become emblematic of Brazil, a Global South giant whose image was tainted by the spectre of death-squads, among other tropicalist and carnivalesque clichés and spectacles (Veissière 2007, 2008). I was aware, then, that innumerable transnational political actors had flocked to Brazil to witness and participate in this politics of street salvation.

What little ethnographic literature existed on the subject of street children in Brazil, in addition, seemed to suggest that this gringo obsession with taking kids off the street was ethnocentrically ingenuous at best, or simply counterproductive. Pointing out that most of the children in the streets of Recife were on average much better nourished and enjoyed more freedom and even happiness than their counterparts living in *favelas* (shantytowns), for example, Tobias Hecht (1998) challenged many of the sensationalist figures that seemed to grossly inflate the number of children actually living in the street, and wondered why so few people were preoccupied with the home children who were left behind in the violent misery of *favelas*. At any rate, if one was concerned with the fate of children in the Global South and Brazil, argued Hecht, the obsession with street kids was largely irrelevant; the main distinction between normal children and those who had been robbed of a childhood, he proposed, should be situated around the idea of nurture. Normal children, he argued, were essentially nurtured beings who were clothed, fed, loved, and cared for by their parents, families, and other cultural, social, and economic structures. Nurturing children, conversely—who, in urban contexts, are seen roaming, working, begging, or living on the streets—have to fend for themselves, and often contribute their meager earnings to feed their families. The point raised by Hecht was that the vast majority of human suffering so many gringos were eager to combat did not take place in the street, but precisely in the subaltern *favela* homes that a relatively small number of children attempted to escape by taking to the street.

The participatory research project I had envisaged before commencing my fieldwork, accordingly, had not been aimed at a potential rescue of children from the street, but on the contrary, an effort to understand and mobilize the spontaneous acts of conscientization that had inspired some children to identify their positions as marginalized and to transcend it through a rebellious life in the street. Before returning to Brazil for a year of fieldwork in Salvador, I had theorized and fantasized at great length about such a participatory ethnography aimed at the mobilization of what I had romantically termed the children's "cartographies of resistance" (Veissière & Diversi, 2008). Once in the field, however, after beginning to grasp the contradictions and pretension of the many agendas of change that were externally imposed on street kids, I found myself compelled to reconsider my position as an activist ethnographer committed to participatory research.

The NGO coordinator's anger at my presence had confronted me with the complex politics of entering the field as a foreigner with a priori ideas about how subaltern Brazilians experienced oppression, doubled by equally a priori agenda about getting them to mobilize against my ideas of oppression. What was the point, it now seemed to me, of rushing into activism and playing Che Guevara when there was so much I needed to learn from the kids themselves? Had not Che himself, as a perpetual outsider, failed in his mission, and ended up causing unnecessary violence and his own death because of his failure to grasp the indigenous visions and priorities of the Bolivian underclass, and because of his insistence on imposing a historically and culturally decontextualized notion of mobilization along Western-industrial constructs of class and proletariat (see McLaren, 2000)?

Gringopolítica as Critical-Consciousness: Border-Thinking and the Path toward an Engaged Ethnography of Resistance

Later, spending more time with poor Brazilians in the street and in their homes (in both *favelas* and less marginal quarters), I began to see further value in my position as a lone ethnographer. As many street children and hustlers became acquainted with my presence as a lone gringo with no particular institutional or proselytizing affiliations, they became more comfortable expressing their cynicism toward the people and institutions intent on rescuing them. I rapidly gained access to stories, which, as other ethnographers in similar contexts had reported before, confirmed that most street kids were very adept at strategically and temporarily embracing the ideological visions of different institutional projects to secure the resources they offered. This also confirmed that far from successfully removing children from the street, such organizations merely provided more resources from which street people can sustain a livelihood (Hecht, 1998; Veissière & Diversi, 2008). Thus, in Salvador, it is not an unlikely scenario to encounter a young boy, who, over the course of a few months, has gone from being an evangelical Christian to obtain food and work from a dogmatic do-gooder, to joining a local African drum-group through an NGO and performed *Candomblé* spirit-possession trances for tourist crowds at local cultural shows, or even in such faraway places as São Paulo or Europe, only to return to begging and hustling in the streets and beaches of Salvador.

The ease with which street kids interacted with me and manipulated my compassion also made me reflect on other aspects—some promising, some perilous—of the implications of being a gringo ethnographer in Salvador. At first it confronted me with the deeply transnational nature of the street as a site of struggle: a site in which, as a particular kind of transnational actor in a transnational picture, I could perhaps have interesting insider's perspectives after all.

Salvador is one of Brazil's most prominent tourist sites, and I rapidly discovered that gringos in general, and the particular assumptions they bring in about the nature of childhood specifically, were important resources among the other means of livelihoods devised by street kids. As I keenly observed the way in which many street kids were particularly adept at producing vulnerable performances of childhood that would strike a chord with tourist sensitivities, I began to sketch important transnational connections in my cartographies of resistance. This realization pointed to new levels of critical consciousness, subaltern knowledges, and resistance on the part of people who were so often assumed to be voiceless victims, and also seemed to place me, as a gringo, at an ideal cultural location in the field. This turned out to be exceptionally promising on both methodological and epistemological levels.

Although I had agonized *ad infinitum* about my problematic position as an outsider before entering the field, I found that being a lone gringo granted me access to a broad variety of strategies, tricks, means of survival and livelihoods, and levels of critical consciousness that would have been hidden from me had I been confined to the moralizing walls of an institutional activist position. My presence as a young gringo fluent in Portuguese in a city globally renowned for sex-tourism and its high number of foreign sexpatriates (Seabrook, 1996), in turn, did not so much constitute an anomaly, as a more or less permanent landmark in a deeply normalized transnational landscape. This is why, after my initial reticence to intrude in the lives of downtrodden Brazilians, I found that my informants did not so much resent as welcome my presence, or at the very least consider it a part of their everyday state of affairs.

This brings me to my epistemological point: what I had considered to be the deeply problematic nature of my gringo positioning, that is, the deeply problematic and violent nature of *gringopolítica* when imposed as an activist agenda, also turned out to be promising on epistemological levels. At its most basic, my own gringo interaction with street kids and other street hookers and hustlers pointed to their acute ability to read global patterns of political economy on the one hand (their knowledge of the incommensurable monetary value and purchasing power of citizens of the North and South, for example), their highly critical awareness of different culturally constructed normative assumptions on the other hand (such as, say, their strategic performances of Modern European constructs of nurtured and vulnerable childhoods), and, in general, to their critical consciousness not only of the local patterns of oppression and hegemony they had clearly rejected, but also of transnational channels of "liberation" through which they could secure more appealing means of livelihoods.

Although this seemed to provide enough material to debunk any possible claims of false consciousness on the part of the oppressed, it also hinted at deeper epistemological possibilities. If, as Diane Nelson (1999) proposed, being a gringo entailed a particular dialogical relationship with Latin America for citizens of the North, it followed that interacting with a gringo as a Latin American entailed a particular dialogical relationship with the North. In this complex space of interaction between gringos and subaltern Brazilians I have called *gringopolítica*, then, lies a dialogically productive zone reminiscent of Gloria Anzáldua's (1987) concept of border-thinking. First articulated from Anzáldua's perspective as a lesbian Chicana situated at the geographical and discursive borderlands between the United States and Mexico, indigenous and *mestiza* identities, Spanish and English languages, and different gendered positions, border-thinking was later developed by Walter Mignolo (2000) as a call to overcome binary oppositions between dominant and dominated forms of knowledges.

Having brought the idea of border-thinking to the attention of the reader—border-thinking on the part of both subaltern people who interact with gringos and the gringos who interact with them—I return to the ethnographic narrative with which I opened this chapter:

> In the silence that follows the weight of Emilio's words, I contemplate in horror the incommensurability of our worldviews. Miles, a Canadian who is also fluent in Portuguese and has, like me, established strong cultural ties with Brazil through marriage and fatherhood, also appears horrified by Emilio's offer. While I remain silent, however, Miles deals with the situation without emphasizing any cultural divide, meeting Emilio halfway. Thanking Emilio for his concern, he guarantees that no harm has been done; that he and Neguinho have had the conversation they needed to have, and that the case has been resolved. "Fine then," Emilio retorts, "have it your way, but you've been warned: you will never get any respect from that little fucker; he will use you and trample upon you, and get everything he can from you, but you will not get any respect from him."
>
> On our way home, my other friend Neil, for whom I have translated the conversation, returns to the question of respect. In the past week, Neil's behaviour has undergone a series of shifts that have caused me to think critically about ethnocentric judgments of people like Emilio. Before the incident with Neguinho, Neil, who does not speak Portuguese and had never been to Latin America before his visit to Salvador, had inadvertently showed me how calloused I had become when he'd expressed concern about a shivering child lying in the street; a child whom I had noticed, but registered as a normal-enough state of affairs. Neil, on the other hand, pointing out that the child was lying in a pool of vomit, was adamant that we should attempt to help him. To make matters worse, the child had been lifted off the ground by a military policeman, and told to scramble off, which he had attempted, staggering, until he'd fallen back on his stomach a mere twenty yards away. After my attempts to rouse him had led to nothing, our feelings of powerlessness had soon given to anger at the general indifference of people around us. Hadn't we, and we alone, been compassionate enough to notice the child? But things, as always, had proved to be more multifaceted. Noticing our distress, a middle-aged man drinking at a nearby bar had approached me and explained that he and his friends, too, had been concerned about the boy, but that they had found that nothing could be done. A woman from the same bar had soon joined in the conversation: "My friends and I talked to the boy earlier too, and offered to call an ambulance for him, but all he wanted was money...money for crack, you know? The whole thing is just a scheme to make you feel sorry for him and give him money." Neil and I had still felt powerless after that, concluding that the child's position was still a violently difficult one, but one to which no short-term band-aid solutions could be applied.

After having been robbed by my friend Neguinho, however, Neil's compassion had subsided, and led to a rapid shift in cultural assumptions: "you know what, Sam? I just don't understand you. We've all been really nice to [Neguinho], have given him money, bought him food and been a friend to him, and this is how he treats us in return? This motherfucker snatched my watch off my wrist last night and would have probably taken more if I hadn't run away, and now he was standing there tonight crying like a baby, and not only do you feel sorry for him, but I get a feeling that you respect him for all this bullshit about rebelling against oppression...and the worst part is...*you* don't get any respect from him either...you're not gonna like this Sam, but I really think that your little friend deserves to get the shit beat out of him."

Later that night, after Neil and I bitterly agreed to disagree, I considered the utter fragility of my position as a gringo ethnographer; one whose worldview, personal relationships, certainties, and even physical safety seem to perpetually collapse as new layers of meaning unfold. Neil had been right, of course, in pointing out my admiration for Neguinho's critical consciousness, border-thinking, and deliberate acts of resistance.

"And why wouldn't I steal?" Neguinho had wailed, "I do this at my own risk knowing I'm gonna get thrashed by the cops, but why wouldn't I do it? As a junkie who's seen as someone who refuses to work—to slave away that is, I can't return to the *favela* where there's a prize on my head. In this city, everybody is trying to get money from the gringos; the cops beat up people like me to clean up the streets so other people can make money from the gringos—but you know what? I want my share too; I want money from the gringos too!"

Now, I am left to wonder how my admiration of Neguinho's critical position also highlights the fragility of my position in so many other moral, physical, and epistemological ways. Reflecting further on my initial desire to avoid this situation, I also consider further fragilities, and, paradoxically, further productive forms of border-thinking.

Secretly, I had also had more reasons to avoid the street in general, and people like Neguinho and Emilio in particular. A week before, a conversation with Emilio, with whom I hadn't spoken before, had revealed the extent to which I had been visible in the neighbourhood: "Oh, I know who you are," he'd proudly claimed, "you've been coming here for over two years. You used to live on [street x] with your pregnant wife, and then you left for a while; now you live in another building on [street y], and you had a birthday party for your kid in the lobby last Sunday."

Later that same week, during the day, Emilio had accosted me. Explaining that he was taking the opportunity to speak to me "without [my] wife around," he had handed me a folded napkin on which the message "call me" had been written, surrounded by little hearts, by one of the waitresses from the British pub. The waitress, with whom I had previously discussed her strategic emphasis of a black *negra* identity, was now eager, it seemed, to continue our conversation in more intimate terms. I had thereafter intended to disengage myself from the street for a while, but had now been confronted with Emilio again. Later still, a conversation with the waitress would reveal even further levels of what can only be called subaltern knowledges and strategies: "Oh, I think you live in that green building, right? On the first floor, isn't it?" she had said. "Well, that's funny, because I worked there as a nanny for three years." After a pause, leaving me to consider the implications of her claim, she had added, "I am still friends with all the doormen."

Conclusion

After ten years of experience in Brazil and two years of fieldwork in Salvador, I am still taken off guard by the myriad ways in which the subaltern understand systems of domination and devise knowledges, networks, and strategies to exercise power within these systems. In Salvador, I have also been particularly impressed with the ways in which oppressed individuals identify both local and transnational dimensions of domination and subsequently seek transnational channels of liberation through their interaction with gringos mediated by a variety of creative tactics (emotional blackmail, theft, seduction, sex, sexual blackmail, etc.).

My encounters with Neguinho, Emilio, and other characters in and around the streets of Salvador have taught me that human resiliency is always stronger than the hegemonies within which subaltern people appear to be captive, and thus, that any intended activism that begins with the premise of false consciousness on the part of the oppressed is naïve at best, violently pretentious at worst, but always displaced and disengaged.

I have argued that participatory activism can only begin with *engaged* forms of theory that can be facilitated through deep interaction with the subaltern resulting in border-thinking for all parties involved. In the case of my engagement in Brazil, I have shown that the tense interactional space between gringos and the subaltern can reveal critical forms of consciousness, and point to acts of resistance that can be identified beyond the discourse of victimization and criminalization.

Yet, I have also shown that this engaged quest for border-thinking necessarily takes the ethnographer to morally and physically dangerous grounds in which one's vulnerability, however, also becomes pregnant with meaning and possibilities.

The danger of engaged theory, I argue, is inevitable.

References

Abdi, A., & Kapoor, D. (Eds.). (2008). *Global perspectives on adult education.* New York: Palgrave MacMillan.

Anzáldua, G. (1987). *Borderlands/la frontera: The new mestiza.* San Francisco: Aunte Lute Press.

Appadurai, A. (1996). *Modernity at large: Cultural dimensions of globalization.* Minneapolis: University of Minnesota Press.

Clifford, J., & Marcus, G. E. (1986). *Writing culture: The poetics and politics of ethnography: A school of American research advanced seminar.* Berkeley: University of California Press.

Ferguson, J. (1994). *The anti-politics machine: "Development", depoliticization, and bureaucratic power in Lesotho.* Minneapolis: University of Minnesota Press.

Freire, P. (2000 [1970]) *Pedagogy of the oppressed.* New York: Continuum.

Hecht, T. (1998). *At home in the street: Street children of Northeast Brazil.* New York: Cambridge University Press.

———. (2006). *After life: An ethnographic novel.* Durham, NC: Duke University Press.

McLaren, P. (1997). *Revolutionary multiculturalism: Pedagogies of dissent for the new millennium.* Boulder, CO: Westview Press.

———. (2000). *Che Guevara, Paulo Freire, and the pedagogy of revolution.* New York: Rowman & Littlefield.

Mignolo, W. (2000). Local histories/global designs: Coloniality, subaltern knowledges, and border thinking. Princeton, NJ: Princeton University Press.

Nelson, D. (1999). *A finger in the wound: Body politics in quincentennial Guatemala.* Berkeley: University of California Press.

Scheper-Hughes, N. (1992). *Death without weeping: The violence of everyday life in Brazil.* Berkeley: University of California Press.

Scheper-Hughes, N., & Sargent C. F. (Eds.). (1998). *Small wars: The cultural politics of childhood.* Berkeley: University of California Press.

Scott, J. C. (1990). *Domination and the arts of resistance: Hidden transcripts.* New Haven, CT: Yale University Press.

Seabrook, J. (1996). *Travels in the skin trade: Tourism and the sex industry.* Ann Arbor: University of Michigan Press.

Veissière, S. (2007). Tropicalism: Misplaced logocentrism and the production of tropical(ist) identities in postcolonial Brazil. In K. Pramela, A. Yaakob, & A. Jaludin (Eds.), *Discourses on culture and identity: An interdisciplinary perspective.* Petaling Jaya: Pearson-Longman.

———. (2008). *Desiring and performing carnivalesque bodies: Emancipation and appropriation in the transatlantic market of sex and culture in Salvador da Bahia, Brazil.* Paper presented at the joint meeting of the Association of Social Anthropologists of the U.K. and the Commonwealth (ASA). "Ownership and Appropriation," University of Auckland, Aotearoa/New Zealand, December 8–12, 2008.

Veissière, S., & Diversi, M. (2008). Popular education, hegemony, and Brazilian street-children: Toward an ethnographic praxis. In A. Abdi, & D. Kapoor (Eds.), *Global perspectives on adult education* (pp. 276–296). New York: Palgrave Macmillan.

CHAPTER SIXTEEN

A Participatory Research Approach to Exploring Social Movement Learning in the Chilean Women's Movement

DONNA M. CHOVANEC &
HÉCTOR M. GONZÁLEZ

Introduction

Along with the struggle in the streets, we had to organize ourselves, to do something more. And we went around adding, one woman, another woman, and over there, quietly, another... We felt afraid, but we met just the same, in the parish, in a house. We talked. We offered ideas. We had to throw out the dictatorship. We began to control the fear. Then came permanent and clandestine meetings. The collective kitchens and the collective buying emerged. Some returned to the militancy of the political party, some had never left it. The organizations: commemorating March 8, MODEMU, CEDEMU, *Mujeres de Luto*, *Mujeres por la Democracia*, the Association of Relatives of the Victims of Repression, the Association of Relatives of the Political Prisoners, the Sebastián Acevedo Movement Against Torture, etcetera.

Women everywhere! Women defending their rights, their lives, their families. The feminine struggle and the feminist struggle. Histories entwined for the common reality.

Democracy in the country and in the home! (And in the political parties too!) We were sharing, learning, educating ourselves and growing...

But the happiness never arrived. On the contrary, the majority of us feel disillusioned, disappointed, hopeless, the pain of seeing that, after many years of the transition period, in many cases, we are in the same situation...

> [But] we have met during this time, we have talked, reflected, analyzed, and shared our concerns, positions, and visions from our own environment, our own reality, our own needs.
>
> We have a great challenge—the future. We are teachers, early childhood educators, community workers, neighbors, and mothers with a tremendous responsibility and commitment to re-educate the next generations. Despite the disappointment and the anger, we are going to build, because we have always done so. There is tremendous potential here, much energy, much strength and courage. It is the time to awaken to action.

These compelling words were translated from a presentation collaboratively prepared by a participatory research team in Chile in July 2002 (Chovanec, Bravo, & González, 2002). In this excerpt, the team describes the courageous experiences of women who fought against the military dictatorship of Augusto Pinochet in Chile during the 1980s and conveys the women's disappointment that the end of the military regime did not bring about the hoped for changes in their lives. At the end, the team acknowledges the women's reflectivity during the research process and issues a rallying cry to action. Significant learning moments that shaped and were shaped by the women's powerful experiences in the women's movement are embedded within the collective narrative from which this segment is drawn. In Foley's (1999) poignant words, "Some of the most powerful learning occurs as people struggle against oppression, as they struggle to make sense of what is happening to them and to work out ways of doing something about it" (pp. 1–2).

In this chapter, we describe an ongoing participatory research project in the small city of Arica in the north of Chile. Working together with activists in the women's movement, represented by local nongovernmental and political organizations, we are researching the women's social and political learning with the hope of catalyzing and extending their learning for mobilization and engagement in the future. First, we provide some background to the research by describing the political-historical context of the community and of the primary organization within which the participatory research process is situated. We follow this with a summary of the participatory research approaches that we have been using, linking it to our study of social movement learning. Finally, we critically reflect on some of the strengths and challenges that have emerged in using these approaches in Arica.

Political-Historical Context

Arica is a city of less than 200,000 inhabitants in the far north of Chile, nearly 2000 kilometers from the country's capital city of Santiago. Arica's geopolitical location as an agricultural, marine, and mining community along the borders with Peru and Bolivia gives it economic and military significance. Politically, Chile enjoyed a longstanding social democratic tradition that, in 1970, culminated in the election of a socialist president, Salvador Allende, in a coalition government. The Popular Unity government immediately enacted policies to reform land ownership, to nationalize resource industries, and to strengthen collective community initiatives. However, supported by the United States, a military coup overthrew the president in 1973 and initiated a regime of terror and economic oppression that lasted for seventeen years. It is now well known that the United States, aided by the internal Chilean elite, installed the military dictatorship to halt the spread of socialism in the region and to experiment with a neoliberal economic model (Chavkin, 1985).

The military regime used both legislative bans and military might to immediately demobilize social and political movements and to reconfigure grassroots and neighborhood organizations. After an initial period of shock and terror, Chileans began to quietly rebuild forces that could collectively resist the military regime. Citizens clandestinely reconstituted political parties rendered illegal by the regime and defiantly established organizations for awareness and action on the plight of political prisoners and missing persons. Communal survival-oriented activities, such as collective kitchens and income-generating initiatives, also emerged in an effort to offset the devastating effects of neoliberal economic policies. After ten years of repression by Pinochet's forces and underground resistance by the once-strong political and social movements, massive public resistance emerged in 1983. The first National Day of Protest occurred on May 11, 1983 followed by the first mass mobilization of the women's movement against the dictatorship on August 11, 1983.

When a broad-based coalition of antidictatorship groups successfully waged a campaign to vote against continued military rule in a 1988 plebiscite, Pinochet was obliged to hold elections in 1989. A coalition of center-left parties, the *Concertación por la Democracia*, initially negotiated the terms of transfer to civil rule and has been successful in each of the four subsequent elections. The terms of the transition included various concessions that would ensure the entrenchment of neoliberalism and the appeasement of social movements (Craske, 1998). Critics have since analyzed the irreconcilable contradictions between the neoliberal economic model and the democratization process in Chile (Bresnahan, 2003).

Despite the political retreat of citizens during the transition period, a new generation of activists has emerged on the political landscape in Chile. Recognizing that the law that governs education, passed just one day before Pinochet surrendered power, establishes an education system that favors wealthier students through decreasing state involvement and increasing privatization, high school students have been mobilizing since 2006 to protest the neoliberalization of education in their country (Chovanec & Benitez, 2008).

The Women's Movement in Arica—CEDEMU and MODEMU

In 1983, the same year that heralded the mobilization of mass protest around the country, a meeting was held that signaled the birth of the women's movement in Arica. The meeting was precipitated by earlier events. One was the detention of four Communist Party male leaders whose wives initiated a series of public protests. The second was the culmination of a year-long clandestine study group attended by women from leftist political parties who had recently become interested in an emerging feminist analysis applied to the military regime in Chile (Kirkwood, 1986). Both sets of women hoped to mobilize a broader base of women to overthrow the dictatorship. In addition, the socialist-feminist contingent hoped to advance the antidictatorship struggle by incorporating a feminist analysis that would simultaneously contribute to women's advancement in Chilean society. However, many women who attended the meeting did not accept this vision.

Ultimately, the meeting resulted in the inauguration of two women's organizations that became the backbone of the women's movement in Arica, *Movimiento por los Derechos de la Mujer* (MODEMU) and *Casa de Estudios de la Mujer* (CEDEMU), whose name was later changed to *Casa de Encuentro de la Mujer*. Working separately and apart, the two organizations engaged in open demonstrations, various clandestine activities and, in the case of CEDEMU, feminist consciousness-raising. Many of these same women simultaneously participated in other elements of the antidictatorship movement, including human rights groups, political parties, communal survival-oriented activities, and initiatives of the Catholic Church (Chovanec, in press). One of the most significant and enduring representations of the women's collaborative work is *Mujeres de Luto* (Women in Mourning). Since 1984, at the height of the repression, the women have stood in black for one hour on the anniversary of the coup (September 11).

Formed in a situation of perceived urgency, the women who aligned with MODEMU protested in the streets by day and engaged in clandestine activities by night—all of which was intended to disrupt the functioning of the regime and to bring attention to the human rights abuses. One MODEMU leader explained, "We wanted women in the struggle, women in the street confronting, fighting, setting up. That's what we wanted, women to shock them." Another added,

> So, we went out, we got together. We weren't fifty or sixty; we were 300, 400 women. We were women from the church, women from the parties, women from the social organizations, neighbors. Singly and clandestinely we invited them. And they came and came and we went out.

As the visible representation of a distinctly feminist movement in Arica, the history of CEDEMU parallels the community-based feminist work documented in Santiago (Valdés & Weinstein, 1993). From her personal experience, a CEDEMU participant highlighted CEDEMU's main endeavors—building self-esteem and independence, supporting women-in-need, and critically integrating their work within the antidictatorship struggle—seen in the following quote.

> One learned to value oneself more, to love oneself more, to move away a bit from the yoke of the house. And apart from this, we did community actions, very nice social actions. We also helped many abused women, pregnant teens,...the women that were [in jail] because they had small children, and [we went] to the hospital as well...Well, apart from this, we also always went downtown to throw pamphlets, we met outside the cathedral to do a minute of silence, to sing the national anthem, to do all those acts against [the regime], to accompany the detained and disappeared...We were always, more than anything, fighting against the dictatorship.

The work of these two women's organizations continued unabated until the end of the dictatorship. However, a number of convergent factors resulted in the pacification of the women's movement in the transition period (Chovanec, in press). One factor was a general assumption that the newly established national ministry for women (SERNAM) would subsume and replace the need for grassroots women's movements. Since the dictatorship, MODEMU has engaged in limited activity in informal political spaces. Yet, to some degree it still holds the power to call together women for specific events such as to celebrate International Women's Day. CEDEMU has persisted as an established but small feminist NGO that focuses on women's rights and issues, such as violence against women and women's health, through a weekly radio show, workshops, personal counseling, research, and advocacy. For example, in recent years, CEDEMU conducted a study and hosted an international conference in Arica on the reproductive health needs of indigenous women in northern Chile and successfully advocated for the first women's shelter in Arica. Of the success of their grassroots work with women, a CEDEMU leader reflected,

> The women will never see things the same way again. I mean, they have the information that allows them to reflect and that allows them to analyze, and that allows them to question...And together with that, incorporating the issues of women, of gender, and...the political consciousness they have. So this is a question that the women—at least the women that passed through CEDEMU, the women that heard our discourse, the women that came to know what we were saying and what we were doing—no longer can say that they don't have the elements to be able to reflect, to be able to analyze the issues.

Although CEDEMU works alongside official government agencies dealing with women's issues and with international funding bodies, the team struggles with numerous challenges well known to the NGO and non-profit sector worldwide: limited funding opportunities, professionalization of the voluntary sector, and enforced neoliberalization and depoliticization of agendas and mandates (Alvarez, 1999; Schild, 1995). Nonetheless, CEDEMU has managed to remain congruent with its socialist-feminist roots, despite the insecurity of this choice. From the beginning, CEDEMU was tenuously connected with the broader feminist movement through its relationship with Movement for the Emancipation of Chilean Women (MEMCH) in Santiago. MEMCH's mission is to promote equal rights and citizenship for women and to strengthen the women's movement in Chile through education in four areas: personal development, social development, labor training, and schooling. Through this network, CEDEMU remains connected to other feminists in Chile and in Latin America and is able to access small sources of funding related to MEMCH's priority areas. The UN Decade for Women provided opportunities for grassroots organizations such as CEDEMU to influence policy documents that contributed to Chile's Equal Opportunity Plan, allowing community organizations to mobilize some services for women using the government's own rhetoric (Franceschet, 2003).

Nevertheless, women activists in Arica are highly dissatisfied with the two decade long "transition" period of Chile's "pseudo-democracy." They recognize the role of neoliberalism in this impasse, citing numerous inadequacies in health, pensions, education, and support for the poor and working classes. They do not feel well served by the official women's ministry, SERNAM, nor has any semblance of the vitality of the earlier grassroots movements reemerged, although both CEDEMU and *Mujeres de Luto* have carried on in Arica (Chovanec, 2009).

Participatory Research with CEDEMU: Exploring Social Movement Learning

In this section, we discuss the processes of initiating and enacting a participatory research project on social movement learning in collaboration with CEDEMU. We first describe how a Chilean-in-exile living in Canada initially proposed the idea for this project, how it was taken up by a Canadian doctoral student and how CEDEMU became integrally involved. Second, we outline the processes of collaborative learning that resulted in the construction of a collective narrative about the women's movement in Arica. Third, we review the process of collaborative analysis, identify the three themes that emerged from the analysis and examine their significance for social movements in Arica and beyond. Finally, we describe the follow-up and extension of this research into new projects.

Grounded in Local Experience, a Research Idea Emerges

Héctor was ten years old on September 11, 1973, the day of the military coup in Chile. He spent his adolescence and young adulthood as a student, youth, and cultural leader in the antidictatorship movement in Arica. Héctor knew most of the women who participated in the diverse facets of the women's movement because they were connected to him on many levels: family, friends, classmates, fellow artists, and political comrades. After many years of grueling resistance work, Héctor arrived in Canada at age twenty-five as a political refugee. Following a period of work with other political refugees, Héctor decided to continue his studies in education.

For many years, Héctor harbored admiration for the bravery of the women, and he often thought that people fighting for justice and human rights around the world ought to know that this small city had given birth to a movement that had withstood fear and severe repression. He thought that their experience could be a learning tool and an inspiration for other movements. When Héctor and Donna met, we were both graduate students at the University of Alberta passionately studying education for social change. Héctor was aware of Donna's interest in women's movements because, at the time, she was collaborating on a research project about antiviolence activists with the members of an abused women's drop in center (Chovanec, 1994). He introduced her to literature and videos about the women's movement in Chile and shared his own experiences as a leader in the antidictatorship movement in Arica.

During a trip to Chile in 1994, Héctor introduced Donna to some of the main protagonists of the women's movement, one of whom was the director of CEDEMU at the time. From this trip, research possibilities for studying the learning dimension of the women's movement in Arica began to take shape. Our exploratory contact with CEDEMU's director was very positive because women were keen to tell the story of the women's movement from their unique perspective far from the capital city.

Constructing a Collective Narrative

In 2001 to 2002, we moved to Arica with our family for nine months to conduct fieldwork in collaboration with CEDEMU. Embedding the research project within CEDEMU made sense because of the organization's longevity and historical connection to the women's movement in Arica and because they had an operational infrastructure to host the project. Despite facing organizational and funding challenges at the time, CEDEMU's team of four women welcomed us warmly and included Donna in the daily functioning and special events of the organization. Our original research ideas were soon modified to more closely align with CEDEMU's expectations and needs. At the first meeting, the CEDEMU team generated a list of women's movement participants to be included in the research, suggested methodological strategies, and clarified their hope that an "outsider's" analysis would shed light on the current impasse in the women's movement, thereby providing impetus for renewed activism in the current era. From this meeting, we formed a research team that included us and two members of CEDEMU, Sandra Bravo and CEDEMU's director, Berta Moreno. The four of us planned the research project together and made continuous modifications based on the response and feedback of other participants. Notably, our initial idea of conducting a small number of in-depth interviews expanded to interviewing any women who were active in the antidictatorship era movement and who were available and wanted to participate. We introduced group sessions as the primary vehicle for including larger numbers of interested women. Over the next six months, we conducted a series of open-ended, dialogical interviews with more than fifty women. In addition, being immersed as a family in the community for an extended period, we had many opportunities for impromptu dialogues with Héctor's friends and fellow activists.

Using this dialogic process, the women openly shared their experiences with us, often building on each others' stories. Even when individually interviewed, the parts of their narratives that intersected with Sandra's or Héctor's became a point of mutuality or convergence. As a stated focus of the research, our dialogues often turned to the many things the women had learned from participating in the women's movement. These included instrumental (e.g., skills), interpretive (e.g., emotional), and critical (e.g., feminist) learnings. Rich learning experiences that were previously submerged were brought to the surface where they could be examined and reactivated for the current situation. The women were sometimes intrigued by the distinctive perspectives that different women assigned to the same events. This was apparent in two early group interviews in which the women recollected and represented their perceptions of the meeting held in 1983 that precipitated the women's movement in Arica. Indeed, this meeting came up in most discussions and became a focal point for the emerging analysis of social movement learning.

Through this process across the many dialogues, the participants were co-constructing knowledge about their shared experiences and building a collective narrative. The experience of interviewing and being interviewed was a collaborative learning experience that prompted revisioning and new understandings of individual and collective experiences of the women's movement in Arica. Thus, while we were all learning about social movement learning, we were also engaging in social movement learning.

The final opportunity for collective reflection on the narrative generated in the interviews occurred through a learning/cultural event organized by the research team. The event, in which we interspersed scripted themes and quotes from the research with music and poetry performed by participants, was a nostalgic celebration of the women's past activism and a call to future action. Through reading aloud the co-constructed narrative prepared by the research team based on the participants' own words, the women relived their shared yet diverse histories, the interrupted and disrupted trajectory of their lives precipitated by the military coup, their many and varied activist roles and activities, their extensive learnings, as well as their disillusionment with the current situation in Chile and their disappointment over the demise of their once-strong movement. We ended the event by collectively denouncing the unremitting injustices in Chile and demanding justice and democracy, thus, prompting the possibility of action emerging from the investigative and educative processes of the research project.

Collectively Theorizing Social Movement Learning

As stated at the outset, how, what, and why activists are learning, educating, creating, and recreating knowledge as they work for social change is the focus of our attention. While in the field, our analysis of social movement learning proceeded alongside the interview process. Often participants introduced their own analysis of the past, present, and future during the individual interviews, groups and informal dialogues. In regularly debriefing formal and informal interviews and in consultations with Berta, key events and concepts emerged early in the research that were useful in theorizing the learning dimension of social movements. Near the end of the fieldwork, the team expanded to include one woman from each of two activist generations to contribute to the collaborative analysis process. Together we tentatively developed three main analytical themes that Donna then elaborated further by engaging the empirical data with the academic literature after leaving the field.

The resulting analysis built on three salient aspects of the women's experience that have implications for understanding the learning that mobilizes social action. First, we learned about the important role of parents and communities in acquiring social and political *consciousness* in the early years, and about the equally important role of political parties, community organizations, liberation theology, or social movement experiences in adolescence or young adulthood. Second, in order for the potential of this consciousness to be fully realized for sustained activism, we learned that systematic and ongoing attention to both elements of *critical praxis* are necessary; that is, action and reflection. In Arica, the urgency of action left little room for critical reflection on the ideological grounding of the movement. Without a critical analysis of the relations of capitalism that installed and sustained the dictatorship and in the absence of a vision of the society they hoped to secure, the women were "unprepared" for the neoliberal "pseudo-democracy" that followed. Third, the younger women activists taught us that participating in social movements had particular *consequences* in personal/interpersonal and social/political lives that left them feeling deeply disillusioned and disconnected from future activism.

Next Steps: Acting on Our Learning

With the addition of two other activists, the research team rearticulated two main purposes when Donna returned to Arica in 2006. First, we disseminated information from the research process to date. We distributed a bound summary of the 2002 project and once again, the team organized a research presentation and cultural event. Using a PowerPoint slideshow with photographic images and text, we collectively presented information from the research analysis interspersed with musical performances. Using the framework developed for the written report, the collective narrative was again presented, this time by a woman from CEDEMU and a woman from MODEMU. We also presented the three themes that arose from the research analysis. As a result, some women gained a fuller understanding of the relationship of learning to social movement activism and were able to connect it to their own experience. However, as we discuss in the next section, the public presentation of the research project (through the document and the event) also reopened previous conflicts between the women's organizations.

The second purpose was to consider the implications of the research findings for the present situation in Chile and for further action and research. This was activated through two group sessions organized by CEDEMU. The discussion at that time prompted reflection on what CEDEMU could do to support the younger women who experienced significant consequences from their social movement experiences during the dictatorship years. The women also noted a growing presence of adolescent women at *Mujeres de Luto*. They speculated on the dynamics and significance of this recent development as well as on their role in "inculcating" political consciousness and activism in the youth.

As is noted in the quote at the start of this chapter, "We are teachers, early childhood educators, community workers, neighbors, and mothers with a tremendous responsibility and commitment to re-educate the next generations." Intersecting the three earlier themes with newer observations, we have initiated an analysis of *intergenerational learning* in social movements that was further stimulated when high school students across Chile mobilized in 2006. Looking now at three generations of activists and the connections between them leads us to consider the learning processes that facilitate activism across generations (Chovanec & Benitez, 2008). To date, we have argued for attention to the critical role of parents and political parties in the political socialization of the next generation. We have also noted the significance of social movement continuity structures that sustain movements during quiescent times through the personal and public actions of longtime and incoming activists (Taylor, 1989; Whittier, 1995). In the case of Arica, women's movement continuity structures are manifested through the enduring presence of CEDEMU and the annually occurring *Mujeres de Luto*. We contend that the current student mobilization has its roots in the political consciousness and praxis of mothers and grandmothers, and in local social movement continuity structures that have quietly sustained an activist culture across generations. The challenge for adult educators is to be actively involved in social movements at the grassroots so as to integrate an intentional pedagogical dimension with the potential to catalyze social activism. This is Donna's task in Canada and the task of Arica activists in their own communities. As CEDEMU's longstanding activist/director made clear, community level actions should be initiated from within.

The visit to Arica in 2006 also resulted in an additional collaboration with a newer community-based NGO, *Centro de Promoción e Integración Sociocultural, Corporación* QUEÑUA. Since then, we have also expanded our Canadian contingent to include a young Chilean-Canadian woman and other academics interested in participatory approaches to studying social movement learning. Together with these newer partners and CEDEMU, we have continued to work across continents to explore the reemergence of the Chilean student movement in light of the findings from the earlier research period. At the time of this writing, we are engaged in a pilot project using participant-generated videos that focuses on the social movement learning of the newest generation of activists in Chile.

Critical Reflection on the Participatory Approach

Participatory research approaches are typically conceptualized as cyclical or spiral processes that simultaneously integrate research, action, and education. By engaging people who are most directly affected by the issue being investigated, participatory research challenges dominant views of research that situate the research process outside the realm of everyday actions and ordinary people. Ideally, the research process is generated by community needs and results in improved circumstances at the local level. For activists and researchers interested in grassroots and systemic change, these approaches have offered a means to engage in a more holistic, inclusive, and empowering approach to research.

The research project described in this chapter is facilitated by a number of attributes of participatory research approaches. Most importantly, the research project and processes were embedded in the community. Himself a member of the community, Héctor provided the impetus for the project, initiated first contact with CEDEMU, and initially activated the networks. The fieldwork was orchestrated in large part by the research team, of which only Donna was not Chilean. Both CEDEMU's and Héctor's reputations and reach into the activist community have been significant in the success of the project. Second, because we lived as a family in Héctor's home community during the fieldwork, we built relationships and connections from an existing history and base of trust, extending it further through our participation in family and community social activities.

Donna's personal and research relationships in the community are built on that foundation and further facilitated by CEDEMU's involvement in the project. Donna was actively involved in CEDEMU's routine of activities and was regularly included in social events involving women in the activist network. The women in the community value the efforts she has made to acquire the language and understand the cultural traditions. In time, she has become a trusted friend and the women have begun to see her as somewhat of an insider rather than as a complete stranger to the movement. As is expected in a Latin American research context (González, 1997), the social and the professional are oftentimes blurred. Third, living for an extended period in a working-class neighborhood kept us more grounded in the daily realities of the community. As neighbors, we observed the contradictions between the widespread discourse of social equality and the material effects of neoliberal socioeconomic policies at the community level.

Notwithstanding the numerous strengths of a participatory approach, many have leveled criticisms at participatory research. One such criticism is that participatory approaches create the illusion of participation while the research process continues to be an imperialist exercise with outside experts manipulating knowledge, ignoring power relationships, and treating groups as homogenous entities (Cooke & Kothari, 2001). In the case of our participatory research project with the women's movement in Arica, this relates in part to the positionality of the outside "expert" researcher. Despite her intimate connections with the community as described earlier, Donna was given certain license to be in that community simply because she was a white North American academic. Although she often felt inadequate in the language and customs of the community, Donna was aware of her privileged status with de facto control over the direction of the project and the final interpretation of the findings. The participants may not have questioned it, but, as researchers, it was incumbent on us to question it: For what purpose am I there? In what ways do I contribute or not to the lives of the people who live in that community?

Related to this are critical questions of representation and voice. Once we presented the findings to the participants and other members of the women's movement, longstanding divisions that existed between groups in the community resurfaced. Some women challenged the written interpretation of their story. They felt unheard, underrepresented, and poorly positioned in the narrative. They felt that the narrative favored the feminist perspective of the women from the host NGO. Thus, our connection with CEDEMU may have created some limitations due to rivalries between different organizations and communities within the women's movement. This reminds us of the problematic ways that communities are defined and bounded. On what basis are particular "voices" representative of particular "communities"? All communities are infused with power based on various factors. In Chile, in addition to gender and power imbalances, powerful affiliations based on political party and social class play a significant role in community politics.

Although the women's response was not entirely unexpected, it prompted us to reevaluate Donna's role and responsibility in and to this community. In a recent conference symposium, she reflects,

> Issues of voice, representation, privilege and power are always foregrounded in feminist research. I asked myself if I had transgressed the feminist code. Where had I gone astray in my careful crafting of the story? *Had* I favoured certain perspectives? Whose voices had I privileged? Was it not my *own* voice that was privileged? I challenged my right to be representing their voice and pronouncing judgments on their lives. (Chovanec, Pauchulo, & Elvy, 2008, p. 418)

Although this kind of self-reflexivity is essential in feminist and participatory research, it can sometimes leave us feeling immobilized, lost, or stuck. In this, Patti Lather (2007) offers theoretical tools for Donna's reflection (page references embedded within the quote refer to Lather):

> Lather asks us to accept that our work is in "ruins" from the start, a symptomatic site of the limits of our knowing...Lather suggests that by chancing any form of representation, I reach the limits of representation. I can never know all the possible permutations and nuances of the "story"—or stories—to be told about the women's movements in Arica, or how many conflicting interpretations might be relished in this space (p. 5). Further, I don't know how my partial and "ruined" representation will be taken up. (Chovanec, Pauchulo, & Elvy, 2008, p. 418)

Lather quotes Lacan in emphasizing the potential of the stuck and lost places in a critical praxis: "Something you don't know anything about allows for hope. It is the sign that you are affected by it" (cited in Lather, 2007, p. 1). Being affected by the women's lives and their learning, and hoping that their sacrifices will contribute to the learning of new generations of those who struggle against injustice in diverse locations is the impetus behind our ongoing participatory research in Arica as we continue to expand the research team and the focus of the research through an ongoing dialogue and smaller scale projects.

Summary and Conclusion

In this chapter, we have provided a descriptive account of an ongoing participatory research project conducted within the women's movement in Arica, Chile. We hope that our description and analysis of this research project provides some insight into the potential of participatory research approaches within diverse locations. To conduct this research, the role of the "insiders," or local people, was essential from the inception of the idea through the fieldwork, analysis, and dissemination to the current pilot project with new methods and emphases. However, we are also challenged to critically reflect on the place of the "outsider's" relationship to the women's realities and knowledges. The researcher's role in facilitating the research process and in collaborating on the educational elements has been more easily identifiable than her responsibility for how the research is taken up in the community, both in terms of reaction and in terms of action. We are left with the hope that our tentative interpretations have meaning and significance in spaces that we cannot predict and that our exploration into intergenerational learning across the three living generations of women activists in Arica will provide useful information for social movements wherever they exist.

References

Alvarez, S. E. (1999). Advocating feminism: The Latin American feminist NGO "boom". *International Feminist Journal of Politics, 1*(2), 181–209.

Bresnahan, R. (2003). Introduction: Chile since 1990: The contradictions of neoliberal democratization. *Latin American Perspectives, 30*(5), 3–15.

Chavkin, S. (1985). *Storm over Chile: The Junta under siege*. Westport, CT: Lawrence Hill.

Chovanec, D. M. (1994). *The experience of consciousness-raising in abused women*. Unpublished master's thesis, University of Alberta, Edmonton, AB.

———. (2009). *Between hope and despair: Women learning politics*. Halifax: Fernwood.

Chovanec, D. M., & Benitez, A. (2008). The penguin revolution in Chile: Exploring intergenerational learning in social movements. *Journal of Contemporary Issue in Education, 3*(1), 39–57.

Chovanec, D. M., Bravo, S., & González, H. (2002). El aprendizaje del poder desde los margenes: Movimientos populares de mujeres en Chile. [Learning power from the margins: Grassroots women's movements in Chile]. Unpublished manuscript.

Chovanec, D. M., Pauchulo, A. L., & Elvy, J. C. (2008). The embracing of difficult knowledge: Locating hope in uneasy spaces. *Proceedings of the Canadian Association for the Study of Adult Education (CASAE) / l'Association Canadienne pour l'Étude de l'Éducation des Adultes (ACÉÉA)*. Vancouver, BC.

Cooke, B., & Kothari, U. (Eds.). (2001). *Participation: The new tyranny?* London: Zed.

Craske, N. (1998). Remasculinisation and the neoliberal state in Latin America. In V. Randall, & G. Waylen (Eds.), *Gender, politics and the state* (pp. 100–120). London: Routledge.

Foley, G. (1999). *Learning in social action: A contribution to understanding informal education*. London: Zed.

Franceschet, S. (2003). "State feminism" and women's movements: The impact of Chile's servicio nacional de la mujer on women's activism. *Latin American Research Review, 38*(1), 9–40.

González, H. M. (1997). *Conciencia socialista and education in Cuba*. Unpublished master's thesis, University of Alberta, Edmonton, AB.

Kirkwood, J. (1986). *Ser política en Chile: Las feministas y los partidos*. Santiago, Chile: FLACSO.

Lather, P. (2007). (Post)Critical feminist methodology: Getting lost [Electronic Version]. *AERA 2007*. Retrieved March 24, 2008 from www.coe.ohio-state.edu/plather/

Schild, V. (1995). NGOs, feminist politics and neo-liberal Latin American state formations: Some lessons from Chile. *Canadian Journal of Development Studies, Special Issue*, 123–147.

Taylor, V. (1989). Social movement continuity: The women's movement in abeyance. *American Sociological Review, 54*(October), 761–775.

Valdés, T., & Weinstein, M. (1993). *Mujeres que sueñan: Las organizaciones de pobladoras en Chile, 1973–1999*. Santiago, Chile: FLACSO-Chile.

Whittier, N. (1995). *Feminist generations: The persistence of the radical women's movement*. Philadelphia, PA: Temple University Press.

CHAPTER SEVENTEEN

Participatory Research, NGOs, and Grassroots Development: Challenges in Rural Bangladesh

BIJOY P. BARUA

Introduction

In development interventions, the terms participatory research, participation, and participatory appraisal are widely used for empowerment and mobilization of marginalized groups in rural Bangladesh. While using these terms, emphasis is generally centered on the active role of Nongovernmental Organizations (NGOs). Over the years, there has been a growing consensus that participatory process can only be attained by the NGOs in rural Bangladesh. More specifically, donors imagine that the NGOs are very innovative, dynamic, flexible, and active in the promotion of democratic development and the establishment of the citizenship rights in the rural society. Furthermore, this initiative has been taken by foreign donors to ensure participatory development due to the failure of conventional development efforts. As a result of the donor's role and support, the NGOs received a total of 379 million dollars through foreign aid that is 34 percent of the total aid flows disbursed to Bangladesh (Transparency International Bangladesh, 2007) in the financial year of 2003–2004. This initiative was taken to implement participatory development projects and programs for the mobilization of the marginalized people in rural Bangladesh. In fact, with the acceptance of participatory grassroots development and foreign aid, the participatory research and grassroots development toward empowerment and liberation of the disadvantaged people has been marginalized (Rahman, 1995). For effective grassroots development, NGOs need to use participatory action research *on continuous basis* within their program operations. I believe that without any spiritual humanism (Sharif, 2004) and sociopolitical, cultural, and ideological commitment one cannot take part in the process of social change for disadvantaged people (Barua, 2001).

This chapter will critically examine the issue of participatory action research and grassroots development in Bangladesh. While examining this issue, I will focus my analysis to the program areas of NGOs and grassroots development in Bangladesh. My proposition is that the NGOs have little time to use participatory research for the empowerment of the rural disadvantaged groups in their program areas since they are occupied with the massive expansion of microcredit programs mainly for their own growth and development. An analysis of this chapter is based on my own experience in program implementation and research involvement in grassroots development with the NGOs in Bangladesh from 1987 to 2007. In this chapter, I will use the terms participatory research and participatory action research interchangeably since these are both used by the NGOs in the implementation of grassroots development in Bangladesh.

Context of NGOs, Politics, and Grassroots Development in Bangladesh

Bangladesh has a total land of 56,977 square miles or 147,570 square kilometers, and a population of more than 158,665 million. Approximately 74 percent of the population lives in rural areas and more than sixty percent lives by agriculture alone (Economic and Social Commission for Asia and the Pacific [ESCAP], 2007). Bangladesh is known as a land of villages with cultural achievement centered around the village life. The precolonial villages were autonomous and self-sufficient "little republics." Political instability rarely disorganized the placidity of their socioeconomic life (Karim, 1996). The colonial power overturned the village communities and their society "by uprooting the native industry" (Marx, 1936, p. 658). The class-based *zamindari* (landlord) system was established in 1793 by the colonial power with the introduction of the Permanent Settlement Act that was designed to institutionalize the private property of land. As a result, the communal ownership disappeared and property rights were given to the men. The colonial power displaced local culture, needs, rights, and local knowledge (Shiva, 1989). More significantly, the people of the country experienced military rule for approximately thirty years between 1947 and 1992. The participation of people in the political process was irregular (Barua, 1999). Postcolonial Bangladesh was heavily dependent on foreign aid, and it received annually less than 2 billion US dollars (Lewis, 2004). Despite this fact, rural people are more than twice as likely to be poor compared to those living in the cities (Saddi, 1998).

Over the years, the people of Bangladesh have struggled for their liberation based on their language and cultural identities (Jahan, 1996). The liberation struggle was inspired by a patriotic song *Jaan debo tabu maan debo na* (we will give up our lives, but we will not give up our chastity or dignity). During the liberation war, millions of people sacrificed their lives for their own culture and language. Despite their long struggle, Bangladesh emerged as an independent nation state through a war of liberation in 1971. This liberation struggle immensely sensitized the young freedom fighters to work for the empowerment of the disadvantaged people. Eventually, these freedom fighters came forward with a commitment to establish NGOs in the postliberation era to help the marginalized people in Bangladesh. The NGOs are viewed as nonprofit civil society organizations that are involved in grassroots promotion for the empowerment of the disadvantaged segment of the population in rural Bangladesh (Barua, 2001). These NGOs determined to regenerate the culture and rural economy to decolonize the minds of the villagers. With this goal and intention, NGOs adopted the Freirian concept to enhance the consciousness of the people and farmers and to sensitize them of their latent potential through the building of grassroots organizations in the 1970s. Most of the NGOs operating in Bangladesh date back to the liberation war and began with relief activities and subsequently moved into development activities. With the changing circumstances, the concept of voluntarism has seen a vertical swing in a direction that includes professionalism, specialization, and formal Western management structures that can be observed in the contemporary NGOs of Bangladesh (Barua, 1999). There are approximately 22,000 NGOs actively involved in the country and of which 2,341[1] receive foreign donation. These NGOs also extended their services to 20–35 percent of the country's population in the areas of education, health, and micro-credit (Lewis, 2004).

Participatory Research and Social Change

Participatory research is an active process in which disadvantaged groups are empowered through collective education and partnership for socioeconomic and political development. Such research is politically committed toward structural social change to dismantle the dominion of the minority group who control the wealth of the society (Brown & Tandon, 1983; Maguire, 1987; Rahman, 1991; Selener, 1997). The participatory research approach allows marginalized people to generate their own knowledge from their daily experiences to liberate them from social oppression. It sensitizes the marginalized people to change their social conditions through a collective effort in their society (Rahman, 1994; Smith, 1997). In the view of Hall (1996)

> Participatory research is a social action process that is biased in favor of dominated, exploited, poor, or otherwise left out people for social change and empowerment. It sees no contradiction between goals of collective empowerment and the deepening of social knowledge. The concern with power and democracy and their interactions are central to participatory research. (p. 887)

In this process of research, the main drive is to promote democratic people's organizations and the restoration of popular knowledge (Rahman, 1991). The main thrust of participatory research is to boost the level of understanding and capacity of the people through collective learning so that they can change their condition through social action within their own sociocultural context. This form of research systematically tries to understand the issues from the perspective of participants. Both the researcher and the people are involved in the process of sharing and learning with the commitment to social change. This process helps both groups at an equal level to build knowledge and act collectively to improve the conditions (Barua, 2002; Rahman, 1994).

Participatory Research and Grassroots Development in Bangladesh

The history of participatory research can be traced back to the late 1970s. Since that time, NGOs began to speak about empowerment, which primarily focused on consciousness raising and the formation of grassroots organizations through popular education programs in the villages of Bangladesh. These popular education programs tended to use the dialogical process based on Freirian thinking. Although facilitating popular education in the rural areas, the NGOs emphasized the active participation of disadvantaged groups who were primarily landless laborers, women, and small farmers, in order to make them conscious and concerned citizens in the rural society (Barua, 2001; Chowdhury, 1989; Hasan, 1983). Participatory research has taken different forms and shapes during the 1970s and 1980s in mobilizing and organizing the rural disadvantaged people in Bangladesh (Barua, 2002; Rahman, 1994). For example, the disadvantaged and powerless rural people participated in the investigation and analysis of the power structure of ten villages with the facilitation of a participatory research study known as The Net and conducted by the Bangladesh Rural Advancement Committee (BRAC), which is a leading NGO in Bangladesh. In this process, the *PROSHIKA-Manabik Unnayan Kendra*, a large NGO in the country, also played an important role to mobilize and empower the landless groups in the rural society (Rahman, 1994). This participatory research specifically reflected how the local elite seized the benefits of the rural poor (Chambers, 1992). This period of time was considered to be a golden age in the implementation of participatory research in Bangladesh. In fact, such trends continued until the late 1980s. With this initial commitment of collective learning and action, NGOs were considered to act as catalysts. Over a period of time, NGOs recognized that

> [d]evelopment must come from the people themselves. People cannot be empowered by the outsiders. Neither the Government, or NGOs

nor political parties can empower the poor of Bangladesh. Therefore, the poor must empower themselves through collective action—in particular through organization and education. (Association of Development Agencies in Bangladesh [ADAB], 1988, p. 8)

Since development strategies over-emphasized economic growth in the late 1980s and early 1990s, the rapid expansion of microcredit programs have become a common trend among the NGOs of Bangladesh. Because of this fact, attention toward participatory research has declined since this process requires long-term social and political commitment and devotion. In the 1990s, NGOs preferred to apply the method of participatory rapid appraisal (PRA) in grassroots development programs in Bangladesh. Consequently, the use of PRA has become the dominant trend among the development actors to ensure the participation of farmers and disadvantaged groups. This PRA is used to improve the efficiency of the project management and is usually carried out in a village within a specific period of time (generally four or five days) by professionals with the cooperation of the farmers, community leaders, and extension workers to collect information and data for project planning and evaluation. Since the emergence of the PRA method in the 1990s, a professional training initiative has entered in the process in order to train and develop the human resources necessary for grassroots development (Thana Cereal Technology Transfer and Identification Project, 2000). Such training has eventually been able to develop and train the required workforce for the facilitation of the PRA method. As a result, the domination of urban educated experts has become a common phenomenon within the grassroots development in rural Bangladesh. This domination has in fact created a kind of social segregation and a culture of binary opposition between the urban experts and rural grassroots workers in the promotion of grassroots development for the disadvantaged people and community. Despite this fact, the issues of participatory research have been marginalized and pushed out of the process of grassroots development (Barua, 2001). Thus the issue of political empowerment through social action has been almost completely ignored through the promotion of *aamar kotha aami koi; tore ektu jigaya loi* (I tell you my own sermon or version; but I ask you for affirmation or verification). This sufficiently indicates that NGOs tend to facilitate this participatory appraisal to justify their actions in the villages through the guided plans and predetermined agenda.

Although the terms of participatory action research and participatory rapid appraisal are used interchangeably in the development programs in Bangladesh, there is a clear distinction between these two. In the view of Chambers (1992),

> [i]n practice, much PRA has similarly been concerned with poverty and equity. But compared with most activist participatory research, PRA has entailed less extended dialogue and those who

have facilitated it have been less ideologically committed *extent* to radical social change. (p. 3)

Although participatory research was initiated with the introduction of Freirian thinking in grassroots development in the later part of the 1970s, this form of research has become a mere catchword in recent times (Barua, 2002; Ribaux & Barua, 1995). Upon review of the NGO activities, three key issues that significantly pose challenges in the promotion of participatory research in grassroots development in Bangladesh were observed.

Competition and Conflict within the NGOs

A climate of competition seems to hinder real cooperation in the development of disadvantaged people within the villages. It has been recognized that "the relations among the large NGOs in Bangladesh have characteristically been distant, even competitive" (ADAB, 1988, p. 1). A serious confrontation and conflict was visible among the leaders of NGOs in the early 1990s and late 2001 within the members of the Association of Development Agencies in Bangladesh (ADAB) even during the time of national political and social crisis in Bangladesh. This situation has gradually led to a climate of conflict among the NGOs. Ribaux and Barua (1995) further illustrated that "a feeling of competitiveness prevails and cooperation between the organizations is rare" (p. 31). Although these NGOs opted for political change and development through collective action by building alliances against oppression in the late 1970s and 1980s, they became virtually inactive in collective action in the 1990s and onward. Despite several efforts to formulate strategies for cooperation and even collaboration, there is still confrontation among the NGOs due to their ideological differences or to their political control over the disadvantaged people in the operational villages. The ideological differences and conflicts among the NGOs became more severe after the national election of 2001. As a result of the ideological conflicts, the ADAB disintegrated in 2001 and formed the Federation of NGOs in Bangladesh (FNB) (Transparency International Bangladesh, 2007). This new organization was created to act as an umbrella agency to reorganize the NGOs in Bangladesh.

Moreover, a bureaucratic framework has gradually been developed within the organizational structures of the NGOs while expanding programs in the country. This situation was particularly reflected in the management of the large NGOs in Bangladesh. Because of this fact, the role of the NGOs has been primarily confined to power and authority rather than collective education. The workshop report of ADAB, Private Rural Initiative Program (PRIP), and Institute for Development Research (IDR) (1992) illustrated that

> [t]hrough examining the clarity of visions, missions and strategies of many NGOs in Bangladesh over time, it is obvious that they have

not gone through an evaluation of their own development process. As a result, they now implement a variety of different activities which are not coordinated, not consistent with each other, not well thought through, and are thus, not as effective as they might be. (p. 2)

Since NGOs have become more concerned about project expansion and money, they have less time to use participatory research in developing their effective programs in the villages for the empowerment of the disadvantaged groups. Interestingly, approximately 60 percent of the newly registered NGOs between 1990 and 1995 were from the Dhaka division even though Dhaka has only a 31 percent share of the country's total population (World Bank, 1996). Obviously, this is because Dhaka is also the center of political and economic power in which NGOs can maintain their network for project funding from the bilateral and international donors. As a result, the power of the NGOs is mainly confined to foreign funding rather than being generated from the people of the villages.

Political Empowerment to Microcredit Programs

Despite their ambivalent role, approximately one thousand[2] NGOs have been involved in the mobilization of the rural disadvantaged and marginalized people in the villages through an implementation of microcredit programs (Afrin, Islam, and Ahmed, 2008; *Palli Karma Sahayak Foundation*, 2007). Their recent attention toward economic development through microcredit has ultimately diverted them from the agenda of political empowerment through collective action. Microcredit has also brought competition and confrontation between the NGOs in the villages. The bigger NGOs were engaged in mobilizing the disadvantaged groups in order to expand their microcredit scheme. In this situation, the middle and small NGOs have had difficulty implementing their development programs for political empowerment through a participatory process. The enormous expansion of programs has also eventually divided the rural people in the villages.

During my fieldwork in the central and southeastern parts of Bangladesh in the winter of 2007, I observed that NGOs were involved in a conflicting situation to gain control over the disadvantaged people through the power of microcredit. The disadvantaged people have been considered as target beneficiaries by the NGOs since they provide them with microcredit. As a result of this, these marginalized people were mainly utilized by the NGOs for their own agenda to attain power and social position. Consequently, the policies intended to offer empowerment to the people through conscientization has turned toward the distribution of money. In fact, "much of their programs have concentrated on basic needs orientation interventions" (Sadeque, 1990, p. 55) rather than political empowerment for social change. In such a situation, the participation of disadvantaged people has often shifted toward money transaction rather than socioeconomic, political, and cultural education. One can assume that this rapid expansion of microcredit programs may decrease the creativity of the people by causing indebtedness to the NGOs. Moreover, it can increase vulnerability and new moral order to patrons who provide the credit. A similar observation is reflected by Chambers, Saxena, and Shah (1989) in the Indian context.

In many cases, such programs are designed through a centralized effort of program planning and do not take into account the local needs of the people. Although the investment in microcredit contributes to provide money to the rural disadvantaged people, it has also become an impediment to collective learning and action for their empowerment and for social change within the villages. Angela, a participant of microcredit program narrated:

> We do not discuss or share our experience in the weekly meeting of the *samity* (grassroots women's organization). We now mainly concentrate our time on credit repayment and deposit collection at the weekly meeting. Only money cannot improve our life. Education is also essential for the socio-economic development and cultural conditions. (quoted in Barua, 1999, p. 67)

These statements indicate that participants of the NGOs programs cannot necessarily organize their education for collective social change through their own grassroots organizations. This is because in most cases the programs are based on generalized needs rather than the needs of the target people. Their participation is mainly confined to loan collection and generation at the weekly meetings with the extension workers. As a result of this process, less priority is given to education, participatory research, and action and more to the contentions with the other NGOs in the villages.

Mobilization, Politics, and the Suspicion of People

The direct involvement of NGOs in the democratic political movement from 1990 to 1991 and 1996 raised some criticism, questioning whether they could and should legally be able to participate in such events. For example, the NGO Affairs Bureau of the Government of Bangladesh stated that NGOs are out of control and do not follow the government's rules and regulations (Holloway, 1998). Although the NGOs have denied that their political roles were similar to the political parties, they welcomed the reinstallation of democracy with the hopes of gaining more freedom in their operation so as to ensure the participation of disadvantaged citizens. The participation of NGOs in the political movements from 1990 to 1991 and 1996 and voter education for the national election in 2001 was not only meant to restore parliamentary democracy for the disadvantaged people, but also to establish their supremacy over the government's administrative structures for their own political gains.[3] Despite their efforts and hopes, the NGOs' activities in regard to corruption and irregularities were questioned by the new democratic government in 1991 and 2001. During that period, questions were raised about the democratic practices of NGOs within their own organizations and in their management structures. As a result, distrust and suspicion as to the role of NGOs developed among liberals and fundamentalists as well as radical groups of the urban educated class.[4] Even the common people of Bangladesh were suspicious about the activities of the NGOs. In the view of some independent radical thinkers, "NGO efforts are never generally directed towards a revolutionary structural change of the society. They rather directly help to maintain the existing socio-economic-political status quo" (cited in Khan, 1990, p. 58). There is growing belief among the people that NGOs fundamentally serve the economic and political agenda of the world capitalist society. They primarily demobilize the rural disadvantaged people from the process of political action and empowerment in the interest of capitalist control. In other words, the NGOs are involved in splitting and alienating the rural people in Bangladesh from the democratic and revolutionary political process with the assistance of Western donors (personal observation, 2007).

There is also a fear of conversion into Christianity since the Western donors fund the NGOs. In Bangladesh, 200,000 mosques, 100,000 *maktabs* (Islamic religious schools) (Shahidullah, 1997), and 5,766 *madrashas* (Islamic schools) (Rashiduzzaman, 1994) are involved in the mobilization of the people through their educational programs. Since the religious institutions are deeply rooted in the sociocultural environment of the village, it is likely impossible to ignore such networks by the NGOs within the villages in Bangladesh (Barua, 2001). On the other hand, NGOs also have their own missions and objectives and "vested interests which may not concur with those of their intended beneficiaries" (Fowler, 1988, p. 13). In many instances, NGOs have been involved in controversial issues related to religion while implementing their programs in the villages. As a result of this, these religious organizations have challenged the NGOs' involvement in programs related to women's education and income generation in the 1990s. Consequently, thousands of education centers of the NGOs were vandalized by religious groups as part of a cultural challenge in Bangladesh. In fact, these religious institutions have a stronghold in determining the sociopolitical behavior in rural life. In Bangladesh, the social and political bonds among the rural people, whether poor or rich, are based on kinship ties and religious affinities. Because of this, even liberal political parties could not avoid the appeal of using socioreligious and cultural values and symbols in order to win the national elections of 1996 and 2001.[5] In Rashiduzzaman's (1994) view, "The NGOs will lose their 'popular base' in the civil society if they become entangled in the religious right vs. liberal controversy in the village" (p. 988). The indigenous social and religious networks act virtually as an informal educational forum for sharing information within the villages. This type of educative sharing takes place spontaneously in the villages through everyday social and emotional relationships and interactions rather than through institutionalized and bureaucratic structures. Incidentally while implementing education for disadvantaged groups in Latin American society, Paulo Freire also could not ignore religious institutions. Rather, he engaged religious institutions to promote education for the benefit of marginalized groups. Freire explicitly mentioned that "My 'meetings' with Marx never suggested to me to stop 'meeting' Christ" (cited in Lange, 1998, p. 92).

Having identified the problems of participatory research in grassroots development within the villages of Bangladesh, I have established that perhaps the mistrust between the NGOs and the people has been developed within the communities because of the NGOs' centralized planning of activities. Despite the limitations and problems within the NGO programs, there is also hope of success in the area of participatory research particularly with the small NGOs. Such success is dependent upon the political, social, and spiritual commitment of the NGOs. This is only possible if their action is taken through collective efforts with the people.

Conclusion

Having analyzed grassroots development in Bangladesh, I have found that the NGOs have dealt and negotiated with various external and internal factors while using participatory action research in their programs. The NGOs have changed their approaches and mandates in order to fit them into the global economic order for the last three and a half decades. The growth of microcredit programs is significant in recent times. Despite this fact, the larger NGOs are basically occupied with the expansion of their programs to consolidate their authority in the villages and to sustain their growth in Bangladesh. They attempt to implement their programs based on generalized needs rather than the specific needs of the rural people. Thus, it is more difficult for them to use participatory action research in their programs due to their long-term commitment and time factors. The management of the larger NGOs is also more centralized in nature. On the other hand, small NGOs are closer to the people and more grounded in the villages. They maintain face-to-face relationships with villagers in order to implement their programs within the villages. Although their decentralized character helps them to be in tune with the needs of the rural people, they also facilitate their programs in the villages in an atmosphere of threat from the bigger NGOs. I believe the rural marginalized and oppressed people can only attain their freedom through collective learning and action and not through service-oriented programs in the villages. Such initiatives can only be achieved if the NGOs' program planning and policies are decentralized and grounded in the villages with local inputs. In other words, participatory research must be linked in all steps of the program for social transformation in the villages.

Notes

I express my sincere thanks to Dr. F. R. Khan, Dr. M. A. Kamal, and Professor E. Haque for their constructive comments during the writing of this chapter. I am also thankful to graduate students of Development Studies (Spring Courses: MDS 503 and MDS 620) for their critical reflections on the issue from the field. My sincere appreciation goes to Mr. Atiqur Rahman Sarker and Mr. M. Shakhaowat Hossain for their time for logistic support and help while writing this chapter.

1. NGO Affairs Bureau, Government of Bangladesh, Dhaka, Bangladesh, April 21, 2008.
2. Grammen Bank first initiated micro credit scheme in 1976 as a pilot project. There are more than three thousand organizations, including NGOs, national commercial banks and specialized financial institutions are implementing micro credit programs in Bangladesh. See Afrin et al., 2008.
3. The author made observations during the period of the 1990s and 2001.
4. In 1999 and 2001, the author gained such experience while conducting field research in the country.
5. The author also witnessed similar practices among the politicians in the country.

References

Association of Development Agencies in Bangladesh [ADAB], Private Rural Initiative Program Trust [PRIP], Institute for Development Research [IDR] (1992). *Vision/mission/strategy notes on workshop on organizational development.* Dhaka: Author.
Association of Development Agencies in Bangladesh [ADAB] (1988). *Future strategies and vision of NGOs.* Draft workshop report, Dhaka: ADAB. Dhaka: Author.
Afrin, S., Islam, N., & Ahmed, S. U. (2008). A multivariate model of micro credit and rural women entrepreneurship development in Bangladesh. *International Journal of Business and Management, 3*(8), 169–185.
Barua, B. P. (1999). *Non-formal education and grassroots development: A case study from rural Bangladesh.* Unpublished master's thesis: Concordia University, Montreal, Quebec, Canada.
———. (2001). *Non-governmental organizations and popular education programs: Can they mobilize culturally appropriate grassroots organizations in rural Bangladesh?* Conference Proceedings, 20th Anniversary Conference of Canadian Association for the Study of Adult Education (pp. 14–19). University of Laval, Quebec, Canada.
———. (2002). *Participatory research, education and rural farmers: A case study from rural Bangladesh.* Conference proceedings, Twenty-second conference of Canadian Association of Adult Education (pp. 19–24). Ontario Institute for Studies in Education, University of Toronto.
Brown, L. D., & Tandon, R. (1983). Ideology and political economy in inquiry: Action research and participatory research. *Journal of Applied Behavioral Science, 19*(3), 277–294.
Chambers, R. (1992). *Rural appraisal: Rapid, relaxed and participatory.* Discussion paper, No. 311. Sussex: Institute of Development Studies.
Chambers, R., Saxena, N. C., & Shah T. (1989). *To the hands of the poor.* London: Intermediate Technology.
Chowdhury, A. N. (1989). *Let grassroots speak: People's participation, self-help groups and NGOs in Bangladesh.* Bangladesh: University Press.
Economic and Social Commission for Asia and the Pacific [ESCAP]. (2007). *Population data sheet.* Bangkok, Thailand: Author.
Fowler, A. (1988). *Non-governmental organizations in Africa: Achieving comparative advantage in relief and micro-development.* Discussion paper, No. 249. Sussex: Institute of Development Studies.
Hall, B. (1996). Participatory research. In A. C. Tuijman (Ed.), *International encyclopedia of adult education and training* (pp. 187–194). Oxford: Pergamon.
Hasan, F. R. M. (1983). Landless mobilization, some conceptual points and the role of the NGO. *Journal of Social Studies, 22,* 129–145.
Holloway, R. (1998). *Supporting citizens' initiatives, Bangladesh's NGOs and society.* Dhaka: University Press.

Jahan, R. (1996). Genocide in Bangladesh (pp. 12–21). In *Bangladesh 1971–1996, people's struggles, 25 years of independence*. Montreal: CERAS.

Karim, N. (1996). *Changing society India, Pakistan and Bangladesh*. Dhaka: Nawraj Kitabasthan.

Khan, T. A. (1990). A compilation of the thinking of "concerned" groups towards the role of NGOs in national development. *ADAB News, Special Issue 2*, May–June, 57–65.

Lange, E. (1998). Fragmented ethics of justice: Freire, liberation theology and pedagogies for the non-poor. *Convergence, XXXI*(1/2), 81–94.

Lewis, D. (2004). On the difficulty of studying "civil society": Reflections on NGOs, state and democracy in Bangladesh. *Contributions to Indian Sociology, No. 38*, 3(24), 299–322.

Maguire, P. (1987). *Doing participatory research: A feminist approach*. Massachusetts: University of Massachusetts.

Marx, K. (1936). The future results of British rule in India. In K. Marx *Selected works, Volume II* (pp. 657–664). Leningrad: Cooperative Publishing Society of Foreign Workers in the U.S.S.R.

Palli Karma Sahayak Foundation. (2007). *Micro-credit financing and poverty alleviation, Bangladesh country report*. Dhaka: *Palli Karma Sahayak* Foundation.

Rahman, A. M. (1991). Theoretical standpoint PAR. In O. Fals-Borda, & A. M. Rahman (Eds.), *Action and knowledge, breaking and the monopoly with participatory action-research* (pp. 13–23). New York: Apex Publishers.

———. (1994). *People's self-development: Perspectives on participatory action research: A journey through experience*. Dhaka: University Press Limited.

———. (1995). Participatory development: Towards liberation or co-optation? In G. Craig and M. Mayo (Eds.), *Community Empowerment: A Reader in participation and development* (pp. 24–33). London & New Jersey: Zed.

Rashiduzzaman, M. (1994). The liberals and the religious right in Bangladesh. *Asian Survey, XXXIV*(11), 974–990.

Ribaux, C. A., & Barua, B. P. (1995). Coaching for organizational development and partnership: An innovative approach. *Medicus Mundi, 58*, 29–45.

Sadeque, S. Z. (1990). NGOs and target groups: Possibilities in the fourth five year plan of Bangladesh (1990–1995). *ADAB News, Special Issue 2*, May–June, 53–56.

Saddi, M. L. K. (1998, February 13). Poverty elevated malnutrition. *Weekly Holiday, Dhaka*, p. 8.

Selener, D. (1997). *Participatory action research and social change*. Ithaca, NY: Cornell University Press.

Sharif, M. (2004). Religious-historical perspective on conflicts and violence: Secular materialism versus spiritual humanism. *International Journal of Sociology and Social Policy, 24*(1/2), 56–84.

Shahidullah, M. (1997). *Swadeshey prahtamik o gono shikkar chalchitra: Visha prathamik o gono shikka andolon* (Bengali), [State of primary and mass education in the country: World primary and mass education movement], Bulletin 26. Dhaka: Bangladesh Community Development Library, p. 206.

Shiva, V. (1989). *Staying alive, women, ecology and development*. New Delhi: Kali for Women.

Smith, S. E. (1997). Deepening participatory action-research. In S. E. Smith, D. G. Williams, & N. A. Johnson (Eds.), *Nurtured by knowledge, learning to do participatory action-research* (pp. 173–263). Ottawa: International Development Research Center.

Thana Cereal Technology Transfer and Identification Project (2000). *Participatory rural appraisal: Practical handbook*. Project publication, no. 24. Dhaka: Author.

Transparency International Bangladesh (2007). *NGOs khate shushannar samasa: Uttaraner uppaya* [Problems of good governance in NGO sector: Way to achieve the goal]. Dhaka: Transparency International Bangladesh.

World Bank. (1996). *Pursuing common goals, strengthening relations between government and development NGOs*. Dhaka: University Press Limited.

CHAPTER EIGHTEEN

Making Space for Youth: iHuman Youth Society and Arts-Based Participatory Research with Street-Involved Youth in Canada

DIANE CONRAD &
WALLIS KENDAL

Research into the experiences of homeless or street-involved youth in Canada (CS/RESORS, 2001) paints a shocking picture of deprivation, lack of support, and victimization suffered by youth on a daily basis. Yet the dominant public perception of street-youth is that they are nuisances at best and dangerous criminals at worst. Rather than providing much needed social assistance for youth caught in dire circumstances, governments cut funding and pass laws that make their survival even more precarious. While further research into youths' street-involvement would add to our understandings of relevant issues, in particular, research is needed that works toward concrete material improvements in the lives of youth. Participatory research that employs youth as co-researchers to investigate youths' street-involvement is a potential vehicle for such engagement.

The study discussed in this chapter partners university-based researchers with ihuman Youth Society, which works with youth in inner-city Edmonton, Alberta, Canada. Using arts-based methods the study aims to empower youth through processes of art making that reflect upon their life experiences, interrogate the social context of their lives, and plant the seeds for personal and social action.

Government Responses to Youth in a Risk Society

With the pressures associated with globalization, today's risk society (Cieslik & Pollock, 2002) is characterized by anxieties and uncertainties over identity and social membership, the detraditionalization of civil society, skepticism over expert knowledge, technological and industrial hazards, economic insecurity, and the unpredictability of the future. Youth are among those most at-risk (Cieslik & Pollock, 2002), or as Giroux (2002) contends, "rather than being *at risk*...youth have become *the risk*" (p. 35).

In the USA's neoliberal social environment, youth are depicted as disposable, a social menace, a generation of suspects, and targets of a war on the young (Giroux, 2002; Pintado-Vertner & Chang, 1999). A moral panic over youth behaviour—concerns over youth dropping out of school, illegal drug use, vandalism, violence, and crime—is fueled by sensationalized media accounts, where youth are "demonized...and derided by politicians looking for quick-fix solutions to crime" (Giroux, 2003, p. 554).

In the criminal justice system and in schools in Canada, American style get-tough-on-crime, zero-tolerance, or three-strikes-you're-out policies are becoming more common (Tibbetts, 2006), with the justice minister calling for trying youth as adults and stiffer sentences (CBC News, 2006).

Only valued as consumers under the hegemonic ideology of postmodern consumerism (Strickland, 2002), youths' use of public space (street corners, shopping malls) for anything other than production and consumption of goods is suspect (Hatty, 1996). Moreover, it is racial minority youth who are the prime suspects (Pintado-Vertner & Chang, 1999)—in Canada these are Aboriginal youth (CS/RESORS, 2001). Young people today are increasingly profiled as deviant, subjected to heightened surveillance, with diminished rights of privacy and personal liberty. "In a society deeply troubled by their presence," says Giroux, "youth prompt in the public imagination, a rhetoric of fear, control, and surveillance" (2003, p. 554).

Rather than being our hope for the future, youth have become scapegoats for society's ills (Giroux, 2003; Strickland, 2002). An erosion of hope on the part of marginalized youth has lead to a culture of refusal (Blake, 2004) characterized by resistance to or nonparticipation in mainstream society.

Within this "youth crisis," youth who are homeless or street-involved are commonly perceived as public nuisances or delinquents—causing problems for ordinary citizens by panhandling and driving away tourists, or dangerous—involved in criminal activity and a threat to public security (Gaetz, 2004). However, they face a myriad of hardships living on the streets and are most often victims themselves (Gaetz, 2004).

Research on homeless youth in Canada since the 1980s (Brannigan & Caputo, 1993; Caputo, Weiler, & Anderson, 1997; CS/RESORS, 2001; Fisher, 1989; Gaetz, 2004; Hagan & McCarthy, 1994; Kufeldt & Nimmo, 1987; McCarthy & Hagan, 1992; Public Health Agency of Canada (PHAC), 2006; Whitbeck & Simons, 1990; Wingert, Higgitt & Ristock, 2005) provides some understanding of the conditions of their lives. Overwhelmingly, the research reports that the majority of runaways who have left family homes, foster care or group homes, have done so because of intolerable living conditions involving poverty, neglect, conflict, violence, substance abuse, and emotional, physical, or sexual abuse from adults in their lives. Other homeless youth are throwaways, kicked out of the house by parents/guardians who are unable (sometimes due to poverty) or unwilling to care for them. This group includes many lesbian, gay, or bisexual youth whose parents do not accept their lifestyles (CS/RESORS, 2001). Aboriginal youth are also overrepresented among homeless youth populations in Canada (PHAC, 2006).

Life on the streets for youth includes multiple hardships: lack of secure housing, poverty due to unemployment or underemployment, hunger, lack of social supports, physical and mental health issues, addiction, victimization including physical violence, theft, and sexual abuse (CS/RESORS, 2001). The day-to-day struggle for survival, for the basic necessities of life including food and secure shelter is ongoing, not allowing street-youth the time and energy required to attain stable housing or the employment skills needed to transition off the streets (Wingert et al., 2005).

The sociopolitical environment hostile to youth compounds the problems for youth who find themselves on the streets. It is adult abuse of children and youth, not adequately sanctioned in our society (Hagan & McCarthy, 1994) that leads many youth to leave home and take to the streets in the first place. Upon leaving home there is little social support available to them. Economic policies since the 1980s have meant cuts to social services, including cuts to child protection services for sixteen- to eighteen-year olds, restrictions on access to income assistance for independently living youth, along with cutbacks in spending on education, health, and housing, that negatively impact the lives of street-youth. Moreover, youth fall between the cracks of the social welfare system which has barriers preventing adolescents from accessing either child or adult services (Fitzgerald, 1995). Few resources exist on the streets that serve the needs of youth specifically (CS/RESORS, 2001; Wingert et al., 2005). Few shelters are exclusively designated for youth, who are reluctant to use adult shelters because they are considered unpleasant, overcrowed and often rife with predators—dangerous places for youth to sleep (Canada Housing and Mortgage Corporation [CMHC], 2001; CS/RESORS, 2001).

Since many street-youth are legally too young to live away from home and too young to work, they are denied legal employment, instead driven to participant in the informal economy or illegal activities to survive. Many municipal governments across Canada have responded to citizens' complaints of the homeless as public nuisances, by banning activities such as panhandling, flagging (holding out signs for money), and squeegeeing (cleaning windshields) (Wingert et al., 2005). In the case of Edmonton, this involves invoking traffic bylaws to curb panhandling and calls for a new bylaw with steep fines to deter coercive begging (Kent, 2007). By making it more difficult for street-youth to meet their basic needs, these policies serve to further marginalize street youth, forcing their speedier entrenchment in street life (Wingert et al., 2005).

Resorting to illegal activities such as prostitution, theft, or drug dealing (McCarthy & Hagan, 1992; Wingert et al., 2005), youth are put at risk of victimization (PHAC, 2006), but will often not seek help from authorities because of their age or involvement in crime (Gaetz, 2004). On the streets then, youth encounter criminal sanction for being homeless, as well as for illegal activities that often follow from homelessness, leaving them in doubly jeopardy (Hagan & McCarthy, 1994). Rather than correctional services, these youth require protection and mental health care (PHAC, 2006 p. 35).

It is not surprising that recent government initiatives to address youth homelessness have proven ineffectual. In 1999, the Canadian government announced the National Homelessness Initiative including increased services and attention to youth homelessness (CMHC, 2001) and the promise of local determination of how objectives would best be met. A follow-up study in Winnipeg, Manitoba (Leo & August, 2005), however, found that the government fell short of its promise, seemingly reluctant to relinquish its power. Priorities identified by community members were largely ignored. In 2006, amid fears of cuts to the Initiative, the government reaffirmed its commitment (Canada News Centre, 2006), yet Canada continues to suffer a homelessness crisis (Laird, 2007).

In Alberta, the Conservative government responded with the intent to create a *ten-year* plan to end homelessness (Government of Alberta, 2007), but like many of its policies, the tightfisted long-range plan will do little to assuage the lives of those suffering poverty and homelessness in the short-term.

Government sponsored research into youth homelessness, including extensive literature reviews (CS/RESORS, 2001; Brannigan & Caputo, 1993; Gratrix, 2005), provide a clear picture of the extent of the problem and the need for intervention, yet effective intervention is slow in coming. Instead, government reviews call for more research to fill the gaps, to amend conceptual and methodological limitations noting difficulties in defining the target population, in estimating its size and characteristics due to street youths' transience and elusiveness. Assessors express dissatisfaction with the randomness of population samples and a resulting lack of generalizability of findings. The reliability of youths' responses is questioned regarding motivations for participating and the accuracy or truthfulness of information provided. Exploratory methods such as descriptive case studies are regarded as not scientifically rigorous enough to provide data needed to inform implementation (Brannigan & Caputo, 1993).

Government studies (Caputo, Weiler & Anderson, 1997; PHAC, 2006), often involving surveys or interviews with street-youth serving as informants in exchange for paltry food vouchers as remuneration, rarely translate into programs with objectives and interventions that match the needs implied by the research (Brannigan & Caputo, 1993; Frankish, Hwang, & Quantz, 2005). Despite ongoing research and program initiatives, homelessness in Edmonton (Edmonton Joint Planning Committee on Housing,

2004; Boyle Street Community Services, 2007), as in other Canadian cities, has steadily increased in recent years, with existing services unable to meet demands. Instead of addressing issues including poverty, abuse, poor access to healthcare, mental illness, and addictions, the moral majority and government criminalize street youth; responses have been "increasingly punitive...while the root causes of youth disaffection and hopelessness [are] ignored" (Strickland, 2002, p. 1).

Participatory Research to Incite Action with ihuman Youth Society

It seems the demands of so-called scientifically rigorous research are problematic or inappropriate for this population, yielding ineffectual results. Quantitative measurement or qualitative description—clearer identification, categorization, or statistical analysis of the characteristics of street-involvement—do little to address the problem. As one youth researcher astutely noted, "A lot of research is used as a stalling tactic in light of the fact that addressing the real problem with real solutions is a daunting task" (De Castell, Jenson, Ibanez, Lawrence, Bennett, Jagosh, et al., 2002, para. 2). Alternative approaches are needed, which address the real material conditions of youths' lives. Positive interventions would include services that promote capacity-building, harm reduction, and alternative educational opportunities (PHAC, 2006).

Participatory research (Fals-Borda & Rahman, 1991; Heron & Reason, 1997; Park, Brydon-Miller, Hall, & Jackson, 1993) confers appropriate purpose to research with marginalized populations. As well as means of producing knowledge, research becomes a tool for community dialogue, education, consciousness-raising, and mobilization for action. As a democratic process, it aims to develop practical knowing in pursuit of worthwhile human purposes and practical solutions to pressing community issues (Reason & Bradbury, 2006). As research for, with, and by the people (Fals-Borda, 1991), a participatory approach engages members of the community as co-researchers rather than as research subjects, ideally, involving participants in all stages of the process—in setting the research agenda, posing questions for inquiry, participating in the collection and analysis of data, and in deciding the outcomes of the process or how the research will be used. Participatory research accentuates the inherent human capacity to create knowledge based on experience—to analyze and reaffirm or criticize popular knowledge, flesh out local problems, examine their contexts, seek and enact solutions (Fals-Borda, 1991).

Participatory research, while gaining a better understanding of the circumstances of youths' street-involvement, can focus on meaningful action. Rather than exploiting youth as informants, research can engage them in producing knowledge and working for change to benefit themselves and other youth. Two exemplars of participatory research with marginalized youth populations involving youth as co-researchers are described here.

1. The Pridehouse Project (DeCastell & Jenson, 2006) partnered researchers in Education at Simon Fraser and York Universities with the community organization Pride Care Society in Vancouver, BC, to engage street-involved queer and questioning youth in an inquiry into their housing and support needs and to establish a basis for fundraising for dedicated housing for this population. The project used multimedia methods for collecting data and disseminating research and included night-time excursions to give out food on the streets.
2. The Mothers on the Move (MOM) Oral History project (Guishard, Fine, Doyle, Jackson, Staten, & Webb, 2005) partnered researchers from the Graduate Center, City University of New York with the South Bronx, NY, activist group MOM, which had for thirteen years been advocating for educational equity for their poor and working-class community. The study engaged community youth in collecting and disseminating oral histories of MOM members while investigating the achievement gap to raise their levels of consciousness around the issue and organizing a youth component to MOM to carry on its advocacy work.

These examples exhibit the potential for participatory research with youth to do something meaningful within the contexts under investigation. Guishard et al. (2005) define their research as "activist youth research" (p. 38). DeCastell et al. (2002) define their participatory research as interventionist research or community-based activist research, "in which you actively set out to DO something in a context in which action is urgent, [rather than] simply 'studying the situation'" (para. 1). They suggest that traditional research approaches can in fact do harm to already vulnerable participants in distracting them from meeting their survival needs or attracting the negative attention of the pimp (para. 1).

The participatory research study described in the following text, still in its community-building and conceptual phases, corresponds with the excellent work ihuman Youth Society has been doing for more than ten years. ihuman offers support to high-risk youth in inner-city Edmonton, many of whom are living and working on the streets. Its stated mission is to promote youths' reintegration into the community through programs involving crisis intervention, life skills development, and arts mentorship (ihuman Youth Society n.d.a, para. 1).

Wallis Kendal, ihuman cofounder and youth outreach worker, describes the sociopolitical/economic situation for youth with whom he works as comparable to a "third world" context in terms of the limited resources and support available to them; he sees many youth caught in a cycle of poverty, despair, and self-abuse with little hope for the future. Government departments have little knowledge of or insights into youths' lives, and in any case, seem more concerned with sanitized facades than real helping. The current trend of increased police presence, security personnel, and new correctional facilities, provides a sense of security as well as employment and revenue to the middle class, while ridding the streets of the poor.

Kendal identifies challenges faced by ihuman youth consistent with issues relevant to street-youth generally. Drug addiction is one of the greatest problems, in particular crystal meth addiction, alongside undiagnosed mental health issues. Other issues of concern include scarcity of resources for programming for high-risk youth; inflexibility of existing programs; racism and discrimination with regards to housing for high-risk youth; difficulty in accessing basic healthcare for youth without official identification; criminalization of youths' addictions; lack of eligibility and access to drug treatment; conflation of mental illness with addictions; physical health concerns related to addiction; assumptions and misunderstandings by the public regarding the needs of addicted youth; profiling and harassment of youth by police and security personnel; growing numbers of twelve- to fourteen-year olds on the street; the tragedy of young mothers and their children living in poverty; and the horrific violence associated with prostitution, drugs, and gangs.

On an average day Kendal meets with or encounters more than twenty young people seeking everything from a place to sleep, an escape from violence or abuse, addiction treatment, food and clothing, health and mental health care, and access to music, art, fashion, or community service. The youth are generally between the ages of twelve and twenty-six and predominately First Nations individuals, with more young Africans, Asians, and Caucasians seeking ihuman's services. The majority of these youth want help, but they are often compromised by their illegal activities directed toward their survival.

The interventions with youth involve staff establishing a support system for each individual including the basic needs of safe shelter, food, mental, and physical health care. Most also need addiction counseling and placement. In this regard, ihuman uses a harm reduction model (MacMaster, 2004), since for many youth, who need longer-term care and constant support to overcome their addictions, more traditional programs are not a good fit. ihuman understands that recovery is an arduous process that takes time and often several attempts. As Kendal explains, "this is not a program where we reshape a person in six weeks" (ihuman Youth Society, n.d.c, para. 5). While the ihuman Youth Society does place emphasis on recovery, it does so in a realistic manner that takes into account the many driving forces behind addiction. A strength-based perspective with an individualized approach is taken for every youth, and if recovery is not an option, safer use is promoted. The average time a youth spends in an ihuman program is three years, during which staff "get kids off the streets, and into arts-based programs which lead them to heal and discover their own identity...while they transition into independent living and steady employment" (ihuman Youth Society, n.d.a, para. 2). During the recovery process ihuman's support and guidance is unconditional; they do not phone police over youths' illegal activities, or censor them in any way. ihuman's success in working with this population is due to the empathetic, gradual, and accepting nature of their programs.

With the aim of creating workable pathways for youth to live healthy lives, the intervention strategies and programs developed by ihuman have had a positive impact for many youth. For Kendal, artist and arts educator, and cofounder Sandra Bromley, also an accomplished artist, the arts are central to ihuman's programming. Arts programming is an effective approach with this population in that the arts are flexible enough to meet the needs of youth on an individual basis. The arts are a vehicle to connect with youth and for youth to explore and express their experiences and understandings of world around them, in ways that accommodate them emotionally, physically and psychically; the arts also accommodate their often turbulent lifestyles. Unlike sports programs that are generally group oriented and make specific demands, arts programs can be individualized. For a recovering youth who does not work well with others, or is not successful at keeping a regular schedule, this is crucial. The majority of programs available for high-risk youth in Edmonton are government funded work-related training programs, more concerned with the needs of the marketplace than the youth, aimed at quickly redirecting youth into the work force. In contrast ihuman's arts programs allow for gradual reintegration, with the well-being of the youth as their primary concern. What the youth need from a program is longevity and the opportunity to connect with a staff member or other youth, to build trust and provide a sense of family.

The ongoing arts-based and life-skills focused educational programs offered by ihuman include

- professional music instruction and production—letting the youth have full creative power of their sonic visions;
- computer competency in music, art, and word-processing programs;
- a visual arts studio;
- an annual youth written and performed theater project with a local theater company;
- a fashion program including public fashion shows;
- guidance with literary development—diary, poetry, storytelling, and spoken word;
- break dancing workshops;
- tutorage and support for academic achievement and literacy; and
- resume, job hunting, and personal coaching.

Following are three success stories about ihuman youth:

1. One young man says, "I went through about 17 different foster homes between the ages of six and ten...I went through a lot of

abuse, physical and psychological. I had a lot of anger issues and I wasn't getting along at school" (ihuman Youth Society, n.d.c, para. 2). Finding himself homeless, he turned to drug dealing to survive. He credits ihuman for providing the means to his recovery. He "began to use the centre's studio to record his rap, it grew to be not only a creative outlet, but also a means to overcome his addiction" (para. 4). Tragically, this young man slipped off track and back into meth use, but ihuman helped him get clean again and into addiction treatment (para. 6).
2. This young woman's story began at age fifteen on the streets. After a year with ihuman she made the decision to turn her life around. She went back to high school and with the help of an ihuman scholarship, studied social work to make a difference and help those with challenges like hers. Employed by ihuman first as a youth worker and now as a professional social worker, she is sought after as a community leader, a speaker, a youth worker, a mentor, and also as a multidisciplinary artist. She now helps high-risk youth across the continent (ihuman Youth Society, n.d.b).
3. At twenty, another young woman discovered ihuman Youth Society, where she began working on plays, poetry, and painting. "At ihuman she found the courage to show her poetry to others who would value it. [For her] writing has always been therapy, a refuge, and a survival tactic" (Sikora, 2005, p. 14). She now works at ihuman helping other youth express themselves through writing and art and is publishing a collection of poetry by inhuman youth.

As is evident from the youths' stories, ihuman believes strongly in youth helping other youth. Graduates from the program, those who ihuman has helped to achieve a level of reintegration, are employed as youth workers and serve as mentors for other youth. ihuman youth give talks and arts-based workshops at conferences, in schools and for community events on drug awareness, addiction, suicide, rebuilding identity, youth leadership, and on how to use the creative arts to make positive change in life (ihuman Youth Society, n.d.a). ihuman has had great success in developing youth leadership and in building a supportive community for youth.

With its aims of peer-to-peer support, consciousness-raising, advocacy and action, ihuman is an ideal partner for participatory research. If more scientifically rigorous evidence were required to support ihuman's capacity for conducting research, a Justice Canada funded feasibility study with ihuman would not have concluded that "it is possible both to engage youth in conflict with the law in artistic endeavors and to have them participate in a research study" (Wright, John, & Sheel, 2005, p. 5).

The proposed participatory research project with ihuman seeks to make use of the ostensible power and legitimacy of university-based research and the academic expertise of university researchers to the benefit of ihuman.

To this end, a number of university-based researchers were recruited from across disciplines relevant to the mission of ihuman, including researchers in Education, Criminology, and Psychiatric Nursing, to collaborate with ihuman youth workers, graduate students, and youth as co-researchers in designing, analyzing, and evaluating the research.

The project's theoretical perspective and methodological design are aligned with ihuman's aims. It takes a participatory approach using arts-based methods (Barone & Eisner, 1997; Conrad, 2004), which corresponds to the integral role of the arts in ihuman's work. As research, the arts, seen as legitimate ways of knowing and making meaning, bring together inquiry and creative processes as powerful alternative means of generating, representing, and sharing knowledge. Community-based, practical, emotional, and embodied ways of knowing are valued. The arts explore questions and express understandings that are not easily accessed or represented through other means in order to more fully understand the nature of human experiences and to contribute to greater quality of life and create a more just and inclusive society.

Striving to end the monopoly of the written word, participatory research commonly incorporates alternative methods such as oral traditions—cultural art forms that are already part of community life such as storytelling, songs, life histories, photography, photo/voice projects, radio, poetry, music, myths, drawing, sculpture, puppetry, and drama. These alternative forms become meeting spaces for cultural exchange (Fals-Borda & Rahman, 1991; DeCastell & Jenson, 2006) offering exciting possibilities for engaging people in expressing and investigating their realities, and for generating knowledge and disseminating research. The arts are a particularly effective means of eliciting responses from groups, including marginalized groups, who do not necessarily concede to or appreciate the dominance of the written word.

Specifically, this project will involve a core group of ihuman youth staff as facilitators, aided by graduate research assistants, working with other ihuman youth, to create digital art video narratives of their life experiences. Digital video technology is a dynamic medium using images and sound to capture and express young people's stories, and a medium for which youth have an affinity. Art videos (which may take the form of visual poems, animations, montages) offer opportunities to capture the complexities of youths' lives and social environments—both the physical realities of their lives and their inner, imagined and felt experiences—to present insights beyond words.

The project is aimed at empowering youth by putting the tools of cultural production into their hands, by actively involving them in advocacy and the search for solutions to youth issues. Engaging youth as co-researchers in a participatory research process will allow them to see themselves as agents of change within their social context. In this way, the project aims to benefit youth participants directly through developing skills, research literacies, and encouraging self-efficacy and empowerment.

As in the Pridehouse project, this project will give youth "an opportunity to train and work with their near peers to develop their own critical accounts of their lives and their futures... and to devise counter-narratives" (DeCastell & Jenson, 2006, p. 242). The youths' art videos will provide insights into their experiences, their readings of the world.

Prospects for Education, Consciousness-raising and Social Change

The Pridehouse and MOM projects illustrate the educational potential and prospects for catalyzing social change through participatory research. In each project, goals for the youth included consciousness-raising, capacity building, skills training, work experience, paid employment, youth leadership development, peer mentoring, and self-advocacy.

For the Pridehouse project, "capacity-building of youth themselves was an explicit goal... to provide youth hired for the project with skills-training, work experience, and paid employment as community-based researchers" (De Castell et al., 2002, para. 1). The youth received training in research methods and media arts; they gained a sense of agency through cultural production and the "development and mobilization of literate competences" (DeCastell & Jenson, 2006, p. 230). Researchers conceptualized the project as both research and pedagogy; through productive activity-based learning and critical thinking they saw "engaged and insightful critique of the daily lives of street-involved queer youth made possible" (p. 242). For youth in the study, "their active design, development and production of a promotional video to fundraise for designated housing helped them access and develop more powerful cognitive, social and political analysis" (p. 243).

For the MOM project (Guishard et al., 2005), with the goal of cultivating critical consciousness among youth, "research training sessions were designed as a democratic environment to empower youth, by expanding and building on their emerging commitments to social change through inquiry" (p. 43). As well as payment and course credit, youth researchers received acknowledgement as authors on publications, gained experience in conducting interviews and participant observation, and had opportunities to present their research. Excerpts from youths' essays about their experiences illustrate their raised awareness; how they came to see themselves as agents for change and as researchers capable of contributing to knowledge production:

1. "I have learned a lot about the activism that goes on within my neighborhood, as well as gained a broader perspective of what has been going on in public schools... I think teaching youths how to be critically conscious and aware of injustices they may not have noticed before is one of the best things that can come of this" (Mothers on the Move, 2006, para. 9).

2. "People who are critically conscious know what is going on and want to make a change. Once they find out they take action" (para. 12).
3. "Through this project I have grown not only into a youth researcher, but an activist. A proud, youth activist" (para. 14).

Guishard et al. (2005) describe the youths' consciousness-raising process in noting that, "as the youth found a vocabulary for naming injustice, they stretched their individual experiences into generalizable collective experiences" (p. 45).

As well as raised awareness for the youth researchers involved in these projects, university-based researchers also learned from the youth. Pridehouse project researchers learned about the risks of homelessness and street involvement for youth of nonhegemonic sexual identification (De Castell et al., 2002, para. 8). MOM project researchers learned "in rich contextual detail, how mothers and youth cultivate an awareness of oppressive educational practices and policies that shape their futures and struggles for justice, against educational inequality" (Guishard et al., 2005, p. 36).

As in these projects, prospects for education, consciousness-raising, and social change through participatory research with ihuman will build on the youths' already emerging awareness of the sociopolitical/economic contexts that shape their life situations. Kendal believes that the prospects for achieving these goals are good, but limited by the youths' day-to-day lived realities, the fact that many are in survival mode and recovering with little energy to spare for critical analysis or action. When they arrive at ihuman the youth already have an embodied sense of how the system (social services, justice, schooling, health care, family) has failed them. On the road to recovery, through dialogue with others, many reach a point of being willing and able to speak out about their experiences, to advocate on their own behalves. Yet, ihuman like other well-intentioned organizations struggling for survival must be guarded in its critique of government from which it also seeks funding. Unlike government programs that demand "deliverables" and number-counting to measure success, ihuman sees success for individual youth in any positive step forward, whether that be walking onto a stage for a fashion show, enrolling in one course at school, recording a song, finding a place to live, selling a painting, or getting off drugs.

With the current pessimistic attitude regarding young people, described as a war on youth (Pintado-Vertner & Chang, 1999), and with homeless and street-involved youth seen as a threat, Blake (2004) claims that the need to develop and acknowledge youths' local literacies is a powerful way to break the culture of refusal and rebuild hope for and with youth. For DeCastell and Jenson (2006), a youth-centered production-based pedagogy, involving cultural production involving multiliteracies, couched in the authoritative discourse of research, provides hope for empowerment, self-understanding, and agency. As Wingert et al. (2005) contend, despite negative portrayals of youth and the real hardships faced by homeless and street-involved youth, research is needed that portrays youth as agents rather than victims, maintains their autonomy; a philosophy of empowerment is critical.

Working with high-risk youth as co-researchers to investigate their life experiences—their struggles as well as the avenues of support available to them—promotes understanding to support the development of appropriate responses to the realities they face. Through participatory activist research, youth can join the discussion of issues concerning them and move toward effecting positive change in their lives.

References

Barone, T., & Eisner, E. (1997). Arts-based educational research. In R. Jaeger (Ed.), *Complementary methods for research in education* (pp. 73–116). Washington: AERA.

Blake, B. (2004). *A culture of refusal.* New York: Peter Lang.

Boyle Street Community Services. (2007). *Winter emergency coordination report.* Edmonton Joint Planning Council.

Brannigan, A., & Caputo, T. (1993). *Studying runaways and street youth in Canada.* Ottawa: Solicitor General Canada.

Canada Mortgage and Housing Corporation [CMHC] (2001). *Environmental scan on youth homelessness.* Retrieved from https://www03.cmhcschl.gc.ca/b2c/b2c/init.do?language=en&shop=Z01EN&areaID=0000000034&productID=00000000340000000010

Canada News Centre. (2006, August 17). *Statement by the honourable Diane Finley.* Retrieved from http://news.gc.ca/web/view/en/index.jsp?articleid=233559&keyword=Statement+by+the+honourable+Diane+Finley.+

Caputo, T., Weiler, R., & Anderson, J. (1997). The street lifestyle study. Ottawa: Alcohol, Drugs and Dependency Issues, Health Canada.

CBC News. (2006, May 26). *Justice Minister outlines initiatives for stiffer sentencing.* Retrieved from http://www.cbc.ca/calgary/story/ca-toews-outline-20060525.html?ref=rss

Cieslik, M., & Pollock, G. (Eds.). (2002). *Young people in a risk society.* Aldershot, UK: Ashgate.

Conrad, D. (2004). Exploring risky youth experiences. *International Journal of Qualitative Methods, 3*(1), http://www.ualberta.ca/~iiqm/backissues/3_1/pdf/conrad.pdf

CS/RESORS. (2001). *Gap analysis of research literature on issues related to street-involved youth.* Ottawa: Research and Statistics, Justice Canada.

DeCastell, S., & Jenson, J. (2006). No place like home. *McGill Journal of Education, 41*(3), 227–247.

De Castell, S., Jenson, J., Ibanez, F., Lawrence, L., Bennett, D., Jagosh, J., et al. (2002). *No place like home: Final research report on the Pridehouse project: People.* Human Resources Development Canada and The PrideCare Society. Retrieved from http://www.sfu.ca/pridehouse/

Edmonton Joint Planning Committee on Housing. (2004). *2003 Edmonton homelessness study.* Retrieved from http://www.moresafehomes.net/Publications/Summary%20Report%202004.pdf

Fals-Borda, O. (1991). Remaking knowledge. In O. Fals-Borda & M. Rahman (Eds.), *Action and knowledge: Breaking the monopoly with participatory action-research* (pp. 146–164). New York: Apex Press.

Fals-Borda, O., & Rahman, M. (Eds.). (1991). *Action and knowledge: Breaking the monopoly with participatory action-research.* New York: Apex Press.

Fisher, J. (1989). *Missing children research report.* Ottawa: Solicitor General Canada.

Fitzgerald, M. (1995). Homeless youths and the child welfare system. *Child Welfare, 74*(3), 717–730.

Frankish, C., Hwang, S., & Quantz, D. (2005). Homelessness and health in Canada. *Canadian Journal of Public Health, 96,* 23–29.

Gaetz, S. (2004). Safe streets for whom? *Canadian Journal of Criminology and Criminal Justice, 46*(4), 423–455.

Giroux, H. (2002). The war on the young. In R. Strickland (Ed.), *Growing up postmodern: Neoliberalism and the war on young* (pp. 35–46). Lanham, MD: Rowman & Littlefield.

Giroux, H. (2003). Racial injustice and disposable youth in the age of zero tolerance. *International Journal of Qualitative Studies in Education, 16*(4), 553–565.

Government of Alberta, (2007, October 29). Province to create a 10-year plan to end homelessness. Retrieved from http://www.alberta.ca/home/NewsFrame.cfm? ReleaseID=/acn/200710/ 22408ED7C28A4-0506-E076-FD9ED5C2CA23289B.html

Gratrix, J. (2005). *Tracking our youth.* Retrieved from http://www.moresafehomes.net/images/ research/Homeless%20Youth%20Lit%20Review%20(OLV).pdf

Guishard, M., Fine, M., Doyle, C., Jackson, J., Staten, T., & Webb, A. (2005). The Bronx on the move. *Journal of Educational & Psychological Consultation, 16*(1/2), 35–54.

Hagan, J., & McCarthy, B. (1994). Double jeopardy. In G. Bridges & M. Myers (Eds.), *Inequality, crime and social control* (pp. 195–211). Boulder, CO: Westview.

Hatty, S. (1996). The violence of displacement. *Violence against Women, 2*(4), 412–428.

Heron, J., & Reason, P. (1997). A participatory inquiry paradigm. *Qualitative Inquiry, 3*(3), 274–294.

ihuman Youth Society (n.d.a). *ihuman.* Retrieved from http://www.ihuman.org

———.(n.d.b). *Real life stories: Demi.* Retrieved from http://www.ihuman.org/ demi.htm

———.(n.d.c). *Real life stories: Martin.* Retrieved from http://www.ihuman.org/martin.htm

Kent, G. (2007, June 20). *Mayor says Saskatoon's approach worth exploring.* Retrieved from http://www. canada.com/edmontonjournal/news/story.html?id=0518f170-45a6-40aa-a0a6-22718c35bcc3

Kufeldt, K., & Nimmo, M. (1987). Kids on the street they have something to say. *Journal of Child Care, 3*(2), 53–61.

Laird, G. (2007). *Shelter: Homelessness in a growth economy. Canada's 21st century paradox.* Retrieved from http://www.ccsd.ca/pubs/2007/upp/SHELTER.pdf

Leo, C., & August, M. (2005). *The federal government and homelessness.* Retrieved from http://www. policyalternatives.ca/index.cfm?act=news&do=Article&call=1082&pA=F2ED34D4&type=5

MacMaster, S. (2004). Harm reduction. *Social Work, 49*(3), 356–363.

McCarthy, B., & Hagan, J. (1992). Surviving on the street. *Journal of Adolescent Research, 7*(4), 412–430.

Park, P., Brydon-Miller, M., Hall, B., & Jackson, T. (Eds.). (1993). *Voices of change.* Westport, CN: Bergin & Garvey.

Pintado-Vertner, R., & Chang, J. (1999). The war on youth. *Colorlines, 7.* Retrieved from http:// www.colorlines.com/article.php?ID=332

Public Health Agency of Canada [PHAC]. (2006). *Street youth in Canada.* Ottawa: Author.

Reason, P., & Bradbury, H. (2006). Inquiry and participation in search of a world worthy of human aspiration. In P. Reason, & H. Bradbury (Eds.), *Handbook of action research* (pp. 1–14). Thousand Oaks, CA: Sage.

Sikora, K. (2005). ihuman's Kristen Sikora. *WWR Magazine, 1*(1), 14–15.

Strickland, R. (2002). What's left of modernity. In Strickland, R. (Ed.), *Growing up postmodern: Neoliberalism and the war on young* (pp. 1–14). Lanham, MD: Rowman & Littlefield.

Tibbetts, J. (2006, September 21). *Government to deliver Canadian version of "three-strikes" law.* Retrieved from http://www.canada.com/vancouversun/news/story.html?id=25505391-0300-446a-ac98-e3987de5678c

Wingert, S., Higgitt, N., & Ristock, J. (2005). Voices from the margins. *Canadian Journal of Urban Research, 14*(1), 54–80.

Whitbeck, L., & Simons, R. (1990). Life on the streets. *Youth and Society, 22*(1), 108–125.

Wright, R., John, L., & Sheel, J. (2005). *Edmonton arts and youth feasibility.* Youth Justice: Justice Canada.

CONTRIBUTORS

Helen Balanoff has lived in the Canadian North for more than thirty years. As an educator she has taught in grade schools, in adult education, and at university and has been a manager in the Department of Education. She has worked on a variety of research projects in the North West Territories (NWT), such as the Legislative Assembly's Special Committee on Education, and the Special Committee on the Review of the NWT Official Languages Act, as well as a number of community-based research projects.

Bijoy P. Barua (PhD, Ontario Institute for Studies in Education, University of Toronto) is an associate professor and chair in the Department of Social Sciences at East West University, Dhaka, Bangladesh. He offers courses at both Graduate and Undergraduate levels in the area of Development Studies, Research Methodology, Civil Society, Ecology, Indigenous Knowledge, Development Management, Gender and Development, and Sociology. He has also worked as researcher, teacher, trainer, program manager, and consultant in Bangladesh, India, Nepal, Sri Lanka, Thailand, Vietnam, Ghana, Switzerland, and Canada. His research interests include international development; comparative education; participatory research; ethnic minorities; indigenous knowledge; engaged Buddhism/ecology; community health; gender and culture development, and civil society. He has published in academic journals such as *International Education* (USA), the *Canadian Journal of Development Studies, Medicus Mundi* (Switzerland), Managerie (Germany), and the *Parkhurst Exchange* (Canadian Medical Educational Journal for Doctors).

Si Belkacem Taieb is first generation of Kabyle born in France. He completed a Bachelors degree in Information and Communication and then moved to Canada in 1998, where he completed a BEd at the University of Alberta. He taught in French immersion and Francophone schools in Alberta and then in the low-income suburbs of Montreal. His next engagement involved an education from a Medicine Man in a First Nations reserve and team-teaching at an Innuit school in the North of Canada. Looking for a culturally equitable school system and inspired by the Medicine Wheel introduced to him by the Medicine Man from Nitassinan (Northern Quebec), he wrote his Master's thesis on "Education as Healing Process" at McGill University (Dean's Honor's list). He is currently working on supporting the building of a Kabyle education system in Tamoult Imazighen (partially in Algeria).

Cynthia M. Chambers is professor of education at the University of Lethbridge and teaches and researches in curriculum studies, language and literacy, and indigenous studies. She received her MA and PhD from the University of Victoria. Her essays are published in edited collections as well as *Canadian Journal of Education, Journal of the Canadian Association of Curriculum Studies, JCT: Journal of Curriculum Theorizing, Educational Insights,* and *English Quarterly.* Along with E. Hasebe-Ludt and C. Leggo, she is author of *Life Writing and Literary Metissage as an Ethos for Our Times* (Peter Lang), a manuscript based on research and writing done in conjunction with SSHRC-funded project on rewriting literacy for cosmopolitan schools. Cynthia lived and worked in the Canadian North for many years. She has experience in community-based research, particularly for the purposes of curriculum that originates from the knowledge base of the community. Besides the SSHRC-funded research with the Ulukhaktokmiut and the North West Territories Literacy Council, she is collaborating with the Kainai (Bloods) on a research project on landscape literacies and the curriculum of place.

Donna M. Chovanec is an assistant professor in educational policy studies at the University of Alberta. She specializes in adult education and learning with a particular interest in social and political learning in social movements. She has worked as a social worker, adult educator, and researcher in a variety of academic, health, and social service settings. For the past several years, she has been engaged in a participatory research project with local NGO's in a small Chilean city with a focus on the learning dimension of the grassroots women's movements in that community. This research has resulted in a number of publications including a forthcoming book from Fernwood entitled *Between Hope and Despair: Women Learning Politics,* two book chapters, an article in the *Journal of Contemporary Issues in Education,* and several papers in refereed conference proceedings.

Diane Conrad is associate professor of drama/theater education at the University of Alberta. Her participatory, arts-based research involves work with high-risk and incarcerated youth. Recent publications include "Exploring Risky Youth Experiences: Popular Theatre as a Participatory Performative Research Method" in the *International Journal of Qualitative Methods;* "Rethinking 'At-risk' in Drama Education: Beyond Prescribed Roles" in *Research in Drama Education;* and a chapter entitled "Participatory Research: An Empowering Methodology with Marginalized Populations" in P. Liamputtong and J. Rumbold's *Knowing Differently: Arts-based and Collaborative Research Methods.*

Gina Currò recently took up the position of language adviser in the academic skills Unit at the University of Melbourne. Her background of experience lies in Applied Linguistics and TESOL; she has taught at Kuwait University's Faculty of Medicine, and has worked in undergraduate, coursework, and higher degree by research settings at James Cook University in Australia and at the Australian National University. Gina enjoys working in cross-cultural literacy, a challenging area of research for students learning how to master discourses in different disciplines.

Héctor M. González is an independent researcher and consultant whose area of specialty is the relationship of education to sociopolitical struggles within the terrain of the prevailing neoliberal system. After spending his youth as a student leader in the antidictatorship movement in Chile, he lived in exile for a number of years in Canada where he obtained a Master's degree in education with a specialization in international education. Currently, he resides in Chile where he continues his lifetime work as an activist-artist (musician and actor) and community-based political organizer. He has assisted in a participatory research project on the women's movement in Chile for several years. From this research, he translated a summary manuscript for distribution in the participating community entitled *Entre la Esperanza y la Desesperanza: El Aprendizaje Social y Político en el Movimiento de Mujeres en Chile.*

Janinka Greenwood is Associate Dean of Postgraduate Studies in Education at the University of Canterbury, New Zealand. A Czech-born New Zealander, her research specialization includes cross-cultural development, research development, and drama as a tool for social change. Current research projects include models of effective tertiary programs for Maori, art as a catalyst for cultural exploration, "Third space" research; building school-based sustainable learning communities; and literacy in the intermediate school.

Steven Jordan is associate professor and chair in the Department of Integrated Studies in Education (DISE), Faculty of Education, McGill University. He is a sociologist of education with research interests in participatory research and evaluation, adult education, education and labor, aboriginal education, and globalization and educational policy making. His articles have been published in journals such as the *British Journal of Sociology of Education, International Journal of Qualitative Studies in Education, Studies in the Education of Adults, Journal of Vocational Education and Training, Comparative Education, Curriculum Studies, Education and Society,* and others. Most recently he has had an entry accepted on PAR in the Sage Encyclopaedia of Qualitative Research and has published book chapters on critical ethnography and globalization. He is a founding director (2001) of Alternative Links, a UK/Canadian NGO aimed at knowledge transfer in the field of information and communications technology (ICT).

Dip Kapoor is associate professor in the Department of Educational Policy Studies, University of Alberta and volunteer research associate of the Center for Research and Development Solidarity (CRDS), an

Adivasi-Dalit peoples' organization in Orissa, India. He is founding member and president of a Canadian voluntary organization that has been engaged with Adivasi-Dalit people in Orissa since 1995. His areas of research include critical sociology of education and development; colonialism, globalization, and education; social movements and popular education/learning; NGO-social movement relations; global education; and qualitative/participatory research. His articles have appeared in *Convergence, Adult Education & Development, New Directions for Adult and Continuing Education, Development in Practice, International Education, Journal of Postcolonial Education,* and *Canadian and International Education.* He is coeditor of *Global Perspectives on Adult Education* (2008, Palgrave Macmillan) and *Beyond Development and Globalization: Social Movement and Critical Perspectives* (University of Ottawa Press, Canada).

Wallis Kendal is a dedicated youth worker. He cofounded and has served as a youth and outreach worker of ihuman Youth Society in Edmonton, Alberta since 1998. Kendal taught art in Edmonton Public Schools and has traveled extensively in third world countries. An accomplished artist, he cocreated The Gun Sculpture in 2000, with Sandra Bromley, which exhibited internationally including in Edmonton and Ottawa, Canada; New York; Seoul, Korea; and Hanover, Germany. Kendal speaks regularly at national and international conferences about youth issues and the arts. In 2000, he won the City of Edmonton Arts Achievement Award; was honored in 2005 as a Time Magazine Canadian Hero; and received the 2006 University of Alberta Alumni Honour Award.

Joe L. Kincheloe is the Canada research chair in Critical Pedagogy in the Faculty of Education at McGill University. He is the author of numerous books and articles about pedagogy, the social construction of knowledge, research, cultural studies, racism, class bias, and sexism, issues of cognition and cultural context, and educational reform. His books include *Teachers as Researchers; Knowledge and Critical Pedagogy: An Introduction, Getting Beyond the Facts: Teaching Social Studies/Social Sciences in the Twenty-first Century; The Critical Pedagogy Primer;* and *Rigour and Complexity in Educational Research: Conceptualizing the Bricolage* (with Kathleen Berry). His coedited works include *White Reign: Deploying Whiteness in America* (with Shirley Steinberg); and the Gustavus Myers Human Rights award winner, *Measured Lies: The Bell Curve Examined* (with Shirley Steinberg and Aaron Gresson). Kincheloe is very concerned with the politics of knowledge and research as they relate to the sociocultural, political, psychological, and educational dimensions of contemporary life.

Valerie Kwai Pun recently earned a Master's degree from McGill University through the Department of Integrated Studies in Education. Her research focuses on the role of adult learning in struggles with development, specifically in Ghana and Africa. She has been working with Wassa Association of Communities Affected by Mining (WACAM) since 2005 and currently resides in Accra, Ghana with her husband and two children.

CONTRIBUTORS

Elizabeth A. Lange is assistant professor of adult education in Educational Policy Studies at the University of Alberta, Canada. Her research focuses on the theory and practice of transformative learning in adult education, adult sustainability and environmental education, participatory action research, and pedagogies for social change. She has almost thirty years of experience as an educator and facilitator in formal and nonformal education settings, particularly community and international contexts. She was honored with the Graduate Research Award by the 37th Adult Education Research Conference for her use of action research in teacher professional development.

Robin McTaggart (BSc / MEd Melbourne, PhD Illinois) is adjunct professor of education at James Cook University, Townsville, Australia. He has a long university career in teaching, research, and management. His academic expertise lies in quality assurance, program evaluation and participatory action research in education, management, and other professions. He has conducted quality assurance audits of several universities, and program evaluations and research studies of action research by educators; discipline-based arts education; arts for disadvantaged youth; instructional computing for intellectually disabled adults; gender equity in private schooling; AIDS/HIV professional development for rural health workers; Aboriginal education in traditionally oriented communities; scientific literacies; distance education provision in technical and further education; and university teaching. He has provided training workshops for several professions in action research and program evaluation in Australia, Canada, Hong Kong, Indonesia, Malaysia, New Zealand, Singapore, Thailand, and the United States.

Christine Hellen Mhina obtained her PhD in International Cultural Studies from the Department of Educational Policy Studies at the University of Alberta. Christine is a participatory action researcher, particularly inspired and informed by the perspectives of international participatory development; community social change; women's agency and activism. Over the last ten years she has studied gender relations and participatory community development and learned the value of including diverse perspectives in action for change. Currently Christine teaches Gender Studies; Aboriginal Society, Social Problems, and Introductory Sociology at Concordia University, College of Alberta; and teaches Swahili language and culture with the Department of Modern Languages and Cultural Studies at the University of Alberta. She is also the director for the diversity program at Sexual Assault Centre of Edmonton.

Edward Shizha is an assistant professor in the Department of Contemporary Studies and Children's Education and Development at Wilfrid Laurier University (Brantford), Canada. His teaching and research interests are in contemporary social problems and education including issues on globalization, postcolonialism, and indigenous education in Africa. Dr. Shizha utilizes participatory action research in his studies. He has published refereed articles that appeared in a number of international journals such as *International Education Journal, The Alberta Journal of Educational Research,* and *Journal of Contemporary Issues in Education,* and book chapters that were published in *Issues in African Education: Sociological Perspectives* (2005), *African Education and Globalization: Critical Perspectives* (2006), and *Global Perspectives on Adult Education* (2008). He is currently working on two book projects, one on Asian and African indigenous knowledge, and a second on current educational challenges in Zimbabwe.

Lynne Harata Te Aika, Kaiarahi Maori College of Education, and the Head of the School of Maori Social and Cultural Studies in Education, University of Canterbury New Zealand, is an educational strategist and planner for *Ngai Tahu Iwi* and a researcher whose specialization includes language revitalization, Maori immersion, and bilingual programs.

Samuel Veissière earned his PhD from McGill University in 2007. He is assistant professor of anthropology and chair of social sciences at the University College of the North in Thompson, Manitoba. He has conducted ethnographic fieldwork on subaltern agency, street livelihoods, and transnational sex and violence in Salvador da Bahia, Brazil. Samuel is currently working on a book entitled, *The Ghosts of Empire: Race, Performance, and Violence in the Transatlantic Cultural Economy of Desire,* which combines his Bahian fieldwork with follow up studies of Brazilian sex-workers in Europe. He is also involved with First Nations Education in Canada/Turtle Island on an ongoing basis.

Shannon Walsh is a filmmaker and writer who splits her time between Canada and South Africa. She is currently working on a feature-length documentary about the human and environmental impacts of the oil sands industry in Alberta. Walsh is a doctoral candidate in the Department of Integrated Studies in Education at McGill University, Montreal.

Cora Weber-Pillwax is associate professor of international and intercultural education at the University of Alberta. She focuses in particular on Indigenous perspectives and experiences, critical and interdisciplinary analysis across the social sciences, participatory action research, and other forms of community driven research as well as First Nations and Métis social and educational issues and policies in Canada. Her publications have appeared in such journals as the *Canadian Journal of Native Education* and the *Journal of Educational Thought.* Currently, she is leading a five year research project: Healing through culture and language: Research with Aboriginal peoples in Northwestern Canada, supported by the Social Sciences and Humanities Research Council of Canada.

INDEX

Aboriginal people(s), 7, 10, 45–46, 54
Aboriginal youth, 252–253
academic research, 32, 42, 50
action research, 16–17, 21, 124–126
  see also educational action research
activism, 29, 116, 156, 166, 173, 177, 211, 216, 221, 230, 231–232
acts of resistance, 211, 220–221
Adivasi, 5
  see also Kondh Adivasi
Adivasi-Dalit, 30, 32, 37
Adivasi-Dalit Ekta Abhijan (ADEA), 29–42
Africa, 4, 139, 141–142, 151, 172
African epistemology, 143
agency, 52, 95, 157, 163, 261
  women's agency, 166
AIDS, 181, 183–186
Alberta (Canada), 251–265
Algeria, 195–207
animism, 197
Arica (Chile), 6, 224–236
arts-based methods, 251
arts-based programs, 7, 257–258
Asia, 4
autonomy, 47, 51, 129–130, 211, 263

Balanoff, Helen, 73, 87
Bangladesh, 239–250
Barua, Bijoy P., 239, 249, 250
Belkacem, Taieb, 195
Berber, 195–204
Berber education, *see* education
border-thinking, 218, 221
Brazil, 189, 209–222
bricolage, 67–68

Casa de Estudios de la Mujer (CEDEMU), 226–230, 232–234

case study, 37, 156, 159, 166, 178, 211
Center for Research and Development Solidarity (CRDS), 32, 36
Chambers, Cynthia M., 73, 87
Chile, 223–237
Chovanec, Donna M., 223, 236,
collaboration, 90, 107, 119, 140, 148
colonialism, 6, 61, 74, 117, 119
  see also neocolonialism
colonization, 7–8, 119, 139, 196, 198, 206
commercialization, 4, 89
Communal Areas Management Programme for Indigenous Resources (CAMPFIRE), 149–151
community, 74, 139, 140, 142–144, 148–150, 164, 177, 233
  community activists, 178–179
  community-based, 33, 79–80, 124, 145, 148, 179, 256, 260
  community development, 166
  community knowledge, 142, 149–150
  community(-ies) of practice, 92–93, 97
  community social action, 125, 129
Conrad, Diane, 251, 263
conscientization, 174, 211, 216
co-optation, 2–3, 6, 16, 212
Cree communities, 47, 51, 55
Cree knowledge, 7
critical complex epistemology, 109–111, 119
critical complexity, 109, 111
Crossmoor (South Africa), 181–193
curriculum, 94, 109, 202
  curriculum internationalization, 90, 92, 99, 101–102
Curró, Gina, 89

data analysis, 33
data collection, 33, 38, 160

Index

development, 9, 21, 31, 37, 66, 68, 124, 141, 150, 156–158, 176, 178–179, 182, 186, 212, 239, 243
   see also community development; grassroots development; international development
disengaged practice, 212

education, 1–3, 39, 40, 55–56, 63–64, 67, 91, 93, 111, 116, 123, 124, 139, 141, 149, 195, 199, 228, 247–248, 255, 262
   Berber education, 202–203
   community education, 176
   education system, 60, 65, 139, 202, 206, 225
   formal education, 4, 148, 150
   informal education, 247
   international education, 94
   popular education, 17, 124, 169, 242
   science education, 146, 151
   sustainability education, 124, 126, 128, 134
educational action research, 126
educational PAR, 3
educational research, 3, 22, 42, 97–100, 107
elders, 33, 65, 69, 76, 81, 159, 165
empowerment, 20, 25, 50–51, 125, 239–240, 242, 245, 262–263
epistemology, *see* African epistemology, critical complex epistemology, Indigenous epistemology(-ies)
ethnography, 213, 216
   critical ethnography, 16, 22–25
   ethnography-in-motion, 181–193
Eurocentric, 108, 139–141, 144, 146

feminist analysis, 226
formal education, *see* education
fundamentalist Islamism, 201
   see also Islam

Ghana, 9, 169–180
Global North, 17, 125
Global South, 4, 15, 17, 22, 124, 210, 215
globalization, 3, 19, 91, 94, 181, 215
González, Héctor M., 223, 236
governmentality, 19, 191
grassroots, 128, 155–156, 227–228
   grassroots development, 239, 243–244, 248

Greenwood, Janinka, 59, 71–72
gringopolítica, 211, 213, 217, 218

hegemony, 3, 42, 108, 123, 147–148, 218
HIV, 183, 186
"Home-less," 190
homelessness, 186, 254
homeless youth, 8, 252–253
human rights, 37, 43, 226, 229

identity, 51, 130, 144, 176, 251
   Amazigh identity, 195–207
ihuman, 7, 256–262
Indigenous communities, 49, 70, 74, 85, 112, 140, 147
Indigenous epistemology(-ies), 74, 147–148, 151, 206
Indigenous knowledge(s), 45–46, 50–51, 139, 141, 143, 144–146, 148–149, 151, 196
Indigenous knowledge systems, 46, 139, 141
Indigenous language(s), 81
Indigenous learning, 149
Indigenous participatory action, 147
Indigenous research methodology (IRM), 45, 53, 57
Indigenous research methods, 46
informal education, *see* education
intellectual imperialism, 85
intentions (of researcher), 54–55, 95, 135, 177
international development, 124–125
international education, *see* education
international students, 91–93
internationalization, 90, 92–95, 101–103
Islam, 197
   see also fundamentalist Islamism

Jordan, Steven, 1, 11, 15, 26

Kabyle diaspora, 197
Kapoor, Dip, 1, 11, 29, 44, 167, 222
Kendal, Wallis, 251
Kincheloe, Joe L., 107, 120–121
knowledge, 96–97
   knowledge production, 109, 119, 142, 145, 147–148, 261
   see also community knowledge

Index

Kondh Adivasi, 29, 31
  see also Adivasi
Kwai Pun, Valerie, 169, 180

land tenure, 155, 166
Lange, Elizabeth A., 123, 136, 250
language immersion program, 60, 62, 68
language revitalization, 60–63, 65, 66, 70
Latin America, 4, 124, 218, 234, 248
learning, 8–9, 16, 19, 22, 24, 30, 38, 56, 70, 131, 141, 161, 169, 174, 177, 224, 232
  collective learning, 159, 242, 246, 248
  informal learning, 8, 9, 24
  see also Indigenous learning; social movement learning
lived-theoretical constructions, 30, 38, 40–41
local knowledge(s), 74, 115, 141

Māori, 8, 59–72, 206
Māori language, 61–64, 68, 70
marae, 59, 63, 65, 206
marginalized groups, 16, 239, 248
marginalized people, 212, 241, 245
margins, 3, 15, 25, 42, 182
McTaggart, Robin, 58, 71, 89, 105
métissage, 75
Mhina, Christine Hellen, 159, 167
micro-credit, 8, 251
microcredit programs, 240, 243, 245–246
mining, 169–180
MOM project, 256, 261–262
Movimiento por los Derechos de la Mujer (MODEMU), 223, 226–227
Mujeres de luto, 226, 232
multinational mining, 169, 171, 173

narrative(s), 33, 41, 51, 129
  collective narrative, 229–230, 232
  video narratives, 7, 260
neocolonialism, 25, 117
  see also colonialism
neoliberal agenda, 19
neoliberal appropriation, 15, 18
neoliberal globalization, 16–17, 31, 43, 129
neoliberal policies, 6, 24, 178, 184, 188
neoliberalism, 18–19, 22, 25, 182–184, 225, 228
nested projects, 67

New Zealand, 8, 59–72, 206
Ngāi Tahu, 61–66, 69
nongovernmental organizations (NGOs), 22, 133, 156, 169, 188, 211, 214, 228, 239–248
Northern Canada, 7, 73–88
Northwest Territories, 73–88

oppressed, the, 16, 30, 119–120, 124, 142, 211, 221
oral traditions, 7, 61, 68, 200, 260
Orissa (India), 5, 29
ownership, 66–67, 74, 83, 148–149

Pākehā, 60, 65–66, 69
Panos, 29, 31
PAR (participatory action research), 5–10, 16–17, 18–19, 22–25, 41, 47–52, 54, 55–57, 59–60, 73–74, 80, 85, 89–90, 93, 95, 103, 107, 134–135, 156, 158, 161, 166, 169–170, 176–179, 248
  epistemological assumptions of, 129–130
  CPAR (community participatory action research), 199, 206
  critical PAR, 109, 111, 113–120
  ontological assumptions of, 128–129
  pedagogical assumptions of, 130–131
  people's PAR, 30–38, 42–43
  reconceptualization of, 123–124, 126, 133
  see also educational PAR
par (participatory academic research), 30–38, 41–42
participatory rapid appraisal (PRA), 243
participatory research, 6, 16–17, 22, 24, 74, 81, 93, 124, 142, 216, 224, 233, 234–236, 239, 241–245, 255–256, 260
pedagogy, 95, 97–100, 130, 133, 135, 261
  critical pedagogy, 110, 112, 190, 211, 213
politicization, 30–31, 33, 41–42
popular education, *see* education
positivism, 23, 107, 109, 113, 117–118
power, 2, 18, 22, 51, 73, 82, 84, 90, 103, 116, 129, 132–134, 191, 212
  asymmetrical relations of, 73, 79
  dominant power, 107–108, 113, 115–116, 118–119
  literacy of, 116
  power relations, 24, 116, 142, 147, 169

practice(s) (educational), 95, 97–100, 102, 124
  subpractices (educational), 97–98
praxis, 4–5, 30, 48, 64, 118, 211–212, 232
  critical praxis, 6, 107, 231, 235
  feminist praxis, 16
Pridehouse project, 256, 261–262
professionalization, 2, 228

qualitative research, 15

research infrastructure, 79–80, 85
resistance, 132, 173, 176–178
  acts of resistance, 211, 220–221
  cartographies of resistance, 216–217
revolution, 49, 53–54
rural communities, 145
rural people, 145, 157, 247–248
rural women, 155–156, 158, 166

Salvador da Bahia (Brazil), 209–222
schooling, 46, 56, 69, 148, 199, 228, 262
science, 108, 111, 113–114, 118, 140–141, 145–148
science education, *see* education
scientistic positivism, 113–114
shack dwellers, 8, 181, 189
shack settlement(s), 181, 183–186, 189
Shizha, Edward, 139, 152–153
social capital, 19–20
social change, 4, 125, 132, 169, 181, 211–213, 239, 241, 246, 261
social justice, 17, 23, 74, 78–79, 82–84, 109, 124
social media, 96–97, 101
social movement learning, 8, 228, 230–231, 233
social movements, 9, 15, 126, 131, 181, 184–185, 186, 211, 225, 231–233
  new social movements, 131, 181, 184
  subaltern social movements, 30, 38, 41
social practices, 96–97, 101
social science(s), 6–7, 15–16, 18, 23, 108, 124, 142
social science research, 16, 73, 77–79, 83, 85–86

social structures, 96–97, 101
social transformation, 131, 133, 211, 248
solidarity, 30, 125–126, 140, 164, 173
South Africa, 181–193
South African Shack Dwellers Organization (SASDO), 189–190
street-involved youth, 7, 251, 262–263
street kids, 211, 215–217
subaltern, 29–31, 37, 43
sustainability development, 8, 127
sustainability education, *see* education

Tanzania, 124, 155–168
Te Aika, Lynne Harata, 59, 72
thematic concern, 89–90
theory, 5, 23, 25, 41, 52, 93, 119, 212,
  complexity theory, 70, 128
  educational theory, 98, 100
  engaged theory, 212, 221
  living systems theory, 128, 131–132, 134
  social theory, 41
  sustainability theory, 123, 126, 128, 134
transformation, 17, 49–50, 52–53, 55, 57, 129
trigger story, 160

Ulukhaktok, 75
Ulukhaktok Literacies Research Project (ULRP), 75–79, 81, 85
Ulukhaktokmiut, 75–76
universalism, 114–115, 117

Veissière, Samuel, 209, 222

WACAM (Wassa Association of Communities Affected by Mining), 8–9, 173–177
Walsh, Shannon, 181, 193
Wassa Association of Communities Affected by Mining (WACAM), 9, 173–178
Weber-Pillwax, Cora, 45, 58, 88
Women's movement (Chilean), 223–237
women's rights, 155, 159, 227
World Bank, 5, 20–22, 25

Zimbabwe, 139–153

GPSR Compliance

The European Union's (EU) General Product Safety Regulation (GPSR) is a set of rules that requires consumer products to be safe and our obligations to ensure this.

If you have any concerns about our products, you can contact us on

ProductSafety@springernature.com

In case Publisher is established outside the EU, the EU authorized representative is:

Springer Nature Customer Service Center GmbH
Europaplatz 3
69115 Heidelberg, Germany

www.ingramcontent.com/pod-product-compliance
Lightning Source LLC
LaVergne TN
LVHW011807060526
838200LV00053B/3689